MODELING AND INTERPRETING
INTERACTIVE HYPOTHESES
IN REGRESSION ANALYSIS

MODELING AND INTERPRETING INTERACTIVE HYPOTHESES IN REGRESSION ANALYSIS

Cindy D. Kam &
Robert J. Franzese Jr.

The University of Michigan Press
Ann Arbor

2010   2009   2008   2007   4   3   2   1

A CIP catalog record for this book is available from the British Library.

Library of Congress Cataloging-in-Publication Data

Kam, Cindy D., 1975–
    Modeling and interpreting interactive hypotheses in regression
analysis / Cindy D. Kam & Robert J. Franzese.
        p.   cm.
    Includes bibliographical references and index.
    ISBN-13: 978-0-472-09969-6 (cloth : alk. paper)
    ISBN-10: 0-472-09969-8 (cloth : alk. paper)
    ISBN-13: 978-0-472-06969-9 (pbk. : alk. paper)
    ISBN-10: 0-472-06969-1 (pbk. : alk. paper)
    1. Regression analysis.   2. Social sciences—Statistical methods.
I. Franzese, Robert J., 1969–   II. Title.

QA278.2.K344   2007
519.5'36—dc22   200605009

    ISBN-13 978-0-472-02299-1 (electronic)

To Anthony & Shelley Kam
—C. D. K.

To Jennifer, Angelina, & Liliana
—R. J. F.

PREFACE

This pedagogical book addresses the modeling, interpreting, testing, and presentation of interactive propositions in regression analysis. We intend it to provide guidance on these issues to advanced undergraduates, graduate students, and researchers in political science and other social-science disciplines. We begin by explaining how verbal statements of interactive arguments and hypotheses translate into mathematical empirical models including, and statistical inferences regarding, interactive terms. The book then provides advice on estimating, interpreting, and presenting the results from such models. It provides next an explanation of some existing general practice rules and, last, a discussion of more advanced topics including nonlinear models and stochastically interactive models. The concluding chapter outlines our general advice for researchers as they formulate, estimate, test, interpret, and present interactive hypotheses in their empirical work.

This project evolved from a previous paper, Cindy D. Kam, Robert J. Franzese, Jr., and Amaney Jamal, "Modeling Interactive Hypotheses and Interpreting Statistical Evidence Regarding Them," presented at the 1999 Annual Meetings of the American Political Science Association. We thank Amaney Jamal for her key role in those origins, and we also gratefully acknowledge Joel Simmons for research assistance in updating some data from the previous project. Finally, we thank Jacob Felson and Michael Robbins for assistance in preparing the final manuscript.

All calculations, tables, and figures can be reproduced using supplementary materials located at www.press.umich.edu/KamFranzese/Interactions.html.

CONTENTS

1. Introduction   1

2. Interactions in Social Science   7

3. Theory to Practice   13
   Specifying Empirical Models to Reflect Interactive Hypotheses
   Interpreting Coefficients from Interactive Models
   Linking Statistical Tests with Interactive Hypotheses
   Presentation of Interactive Effects

4. The Meaning, Use, and Abuse of Some Common
   General-Practice Rules   93
   Colinearity and Mean-Centering the Components of
      Interaction Terms
   Including x and z when xz Appears

5. Extensions   103
   Separate-Sample versus Pooled-Sample Estimation of
      Interactive Effects
   Nonlinear Models
   Random-Effects Models and Hierarchical Models

6. Summary   131

   Appendix A. Differentiation Rules   133

   Appendix B. Stata Syntax   136
   Marginal Effects, Standard Errors, and Confidence Intervals
   Predicted Values, Standard Errors, and Confidence Intervals
   Marginal Effects, Using "lincom"

   References   147

   Index   151

I

≈≈≈

INTRODUCTION

Social scientists study complex phenomena. This complexity requires a wide variety of quantitative methods and tools for empirical analyses. Often, however, social scientists might begin with interest in identifying the simple impact of some variable(s), X, on some dependent variable, Y. Political scientists might study the effect of socioeconomic status on an individual's level of political participation or the effect of partisanship on a legislator's voting behavior. Scholars of comparative politics might be interested in the effect of electoral rules such as multimember versus single-member districts on the party composition of legislatures. Scholars of international relations might study the effect of casualties on the duration of military conflict. Psychologists might study the effect of personality traits on an individual's willingness to obey authority or the effect of an experimental manipulation of background noise on an individual's ability to solve a problem. Economists might investigate the effect of education on labor-market earnings or the effect of fiscal policy on macroeconomic growth. Sociologists might examine the effect of the number of years an immigrant has lived in a host country on his or her level of cultural and linguistic assimilation. Each of these examples posits a simple relationship between some independent variable and a dependent variable.

One of the simplest empirical model specifications for these types of queries is the linear-additive model. The linear-additive model proposes that a dependent variable has a linear-additive, that is, a simple, constant, unconditional, relationship with a set of independent variables. For each unit increase in an independent variable, the linear-additive model assumes that the dependent variable responds in the same way, under any conditions. Much of the quantitative analysis in print across the social sciences exemplifies this approach.

Such linear-additive approaches address what might be described as a "first generation" question, where researchers seek to establish whether some relationship exists between an independent variable, X, and a dependent variable, Y. A "second generation" question adds an additional layer of complexity, asking not simply whether some relationship exists between an independent variable and a dependent variable but under what conditions and in what manner such a relationship exists: for example, under what conditions is the relationship greater or lesser? Thus, this slightly more complex question posits that the effect of some variable, X, on the dependent variable, Y, depends upon a third (set of) independent variable(s), Z.[1]

One could imagine adding such a layer of complexity to each of the preceding examples. For example, the political scientist studying the effect of socioeconomic status on political participation might suspect that this effect depends upon the level of party mobilization in an election—the participatory gains from socioeconomic status might be attenuated when political parties do more to mobilize citizens at all levels. The effect of a legislator's partisanship on his or her votes surely depends upon whether bills have bipartisan or partisan sponsorship. The effect of multimember districts on the party composition of legislatures likely depends on a third variable, societal fragmentation. The effect of casualties on the duration of military conflict might depend on domestic economic conditions. The psychologists might expect the effects of certain personality traits on individuals' willingness to obey authority to increase, and of others to decrease, with age, and the effect of background noise on problem-solving ability might depend on how well rested the subject is. The economist studying the returns to education might expect booming macroeconomic conditions to magnify, and slumping ones to dampen, the effect of education on labor-market earnings; and the one studying fiscal policy would predict zero real-growth effects when the public expected policies and nonzero effects only when policies were unexpected. Finally, the sociologist studying immigrant assimilation might expect the years lived in the host country to have a greater effect for immigrants from source countries with smaller diasporas than for immigrants from source countries with

---

1. For expositional ease and clarity, the discussion that follows primarily focuses on a single variable, x, and a single variable, z, as they relate to a single dependent variable, y. The general claims extend naturally to vectors X, Z, and Y.

larger diasporas, the former perhaps being forced to assimilate more quickly. Social scientists often evaluate such hypotheses using the linear-interactive, or multiplicative, term.[2]

Interaction terms are hardly new to social-science research; indeed, their use is now almost common. Given the growing attention to the roles of institutions and institutional contexts in politics, economics, and society, and the growing attention to how context more generally (e.g., information environments, neighborhood composition, social networks) conditions the influence of individual-level characteristics on behavior and attitudes, interactive hypotheses should perhaps become even more common. However, despite occasional constructive pedagogical treatises on interaction usage in the past, a commonly known, accepted, and followed methodology for using and interpreting interaction terms continues to elude social scientists. Partly as a consequence, misinterpretation and substantive and statistical confusion remain rife. Sadly, Friedrich's (1982) summary of the state of affairs could still serve today:

> while multiplicative terms are widely identified as a way to assess interaction in data, the extant literature is short on advice about how to interpret their results and long on caveats and disclaimers regarding their use. (798)

This book seeks to redress this and related persistent needs. Our discussion assumes working knowledge of the linear-additive regression model.[3] Chapter 2 begins our discussion of modeling and interpreting

---

2. Scholars also refer to the *interactive term* as the *multiplicative* or *product term*, or the *moderator variable*, depending on the discipline. We use *interactive term* and *multiplicative term* interchangeably. In the field of psychology, distinctions are made between *mediator* and *moderator* variables (Baron and Kenny 1986). The distinction is similar to that made in other disciplines, including sometimes in political science, between *intervening* and *interactive* variables, but this terminology is not consistently applied across disciplines and sometimes not even within disciplines. Our discussion applies to *moderator* and *interactive* variables, which Baron and Kenny (1986) define as "a qualitative . . . or quantitative . . . variable that affects the direction and/or strength of the relation between an independent or predictor variable and a dependent or criterion variable" (1174). We reiterate that interactive terms apply when scholars theorize that z affects the *existence or magnitude of the relationship* between x and y, not when scholars believe that some variable z affects the *level* of some variable x that in turn relates to y. This latter argument represents z as a mediating or intervening variable, and an interaction term is not the appropriate way to model it. Instead, mediation is more appropriately modeled by linear-additive regression in various sorts of path analysis; moderation implies interactions.

3. For a refresher on the linear-additive regression model, the interested reader might consult Achen (1982).

interactive hypotheses. This chapter emphasizes how interactive terms are essential for testing common and important classes of theories in social science and provides several theoretical examples in this regard.

In chapter 3, we offer advice on connecting theoretical propositions that suggest interactive relationships to empirical models that enable the researcher to test those interactive hypotheses. We then show which standard statistical tests (certain common t- and F-tests) speak to which of the specific hypotheses that are typically nested in interactive arguments. We discuss a generic approach to interpreting the estimation results of interactive models and illustrate its application across an array of different types of interactive relationships where different types and numbers of variables are involved. We also address the presentation of interaction effects. In all cases, we urge researchers to go beyond merely reporting individual *coefficients* and standard-error estimates. Instead, we strongly suggest graphical or tabular presentation of results, including *effect-line* graphs or *conditional-coefficient* tables, complete with standard errors, confidence intervals, or significance levels of those *effects* or *conditional coefficients*. We discuss and provide examples of several types of graphs that facilitate interpretation of interaction effects, including effect-line plots, scatter plots, and box plots. We also provide instructions on how to construct these plots and tables with statistical software commonly used in social science, in addition to specific mathematical formulas for their elements. Our approach underscores the importance of understanding the elementary logic and mathematics underlying models that use interactive terms, rather than simply providing a set of commands for the user to enter mechanically. If students and scholars understand the foundations of this generic approach, then they will be well equipped to apply and extend it to any new theoretical problems and empirical analyses.

In chapter 4, we consider certain general-practice rules for modeling interactions that some previous methodological treatments advise and social scientists often follow. We suggest that some scholars may be misinterpreting these rules, and we argue that such general rules should never substitute for a solid understanding of the simple mathematical structure of interaction terms. For example, "centering" the variables to be interacted, as several methods texts advise, alters nothing important statistically and nothing at all substantively. Furthermore, the common admonition that one must include both x and z if the model contains an xz term is an often-advisable philosophy-of-science guideline—as an application of Occam's razor (that the simplest explanation is to be preferred) and, as a practical matter, such inclusion is usually a much safer adage than exclusion—but it is neither logically nor statistically necessary and not always advisable, much less required.

Chapter 5 discusses some more technical concerns often expressed regarding interactive models. First, we discuss the question of pooled-sample versus separate-sample estimation that arises in every social-science discipline. We show that estimating interactive effects in separate samples is essentially equivalent to estimating them in a pooled sample but that pooled-sample estimation is more flexible and facilitates statistical comparisons even if one might prefer separate-sample estimation for convenience in preliminary analyses. The chapter then discusses nonlinear models. Although all of our preceding discussion addresses multiplicative terms exclusively in the context of linear-regression models, statistical research in social science increasingly employs qualitative or limited dependent-variable models or other models beyond linear ones. We show first that most of the discussion regarding linear-regression models holds for nonlinear models, and then we provide specific guidance for the special case of interactive terms in two commonly used nonlinear models: probit and logit. Finally, we address random-coefficient and hierarchical models. As Western (1998) notes, using multiplicative terms alone to capture the dependence on z of x's effect on y (and vice versa) implicitly assumes that the dependence is deterministic. Yet this dependence is surely as stochastic as any other empirical relationship we might posit in social science, and so we should perhaps model it as such. Many researchers take this need to incorporate a stochastic element as demanding the use of random-coefficient models. Others go further to claim that cross-level interaction terms—that is, those involving variables at a microlevel (e.g., individual characteristics in a survey) and at a more macrolevel (e.g., characteristics of that individual's state of residence)—that do not allow such stochastic elements may be biased. As a consequence, a growing number of scholars recommend the use of hierarchical linear models (HLM) or first-stage separate-sample estimation of microlevel factors' effects followed by second-stage estimation of macrolevel and macrolevel-conditional effects from the first-stage estimates. Actually, separate-sample versus pooled-sample estimation and whether one must apply two-stage or HLM techniques in multilevel data are related issues, and, as we show, under some conditions, the simple multiplicative term sacrifices little relative to these more complicated approaches. Moreover, steps of intermediate complexity can allay those concerns (not quite fully, but likely sufficiently) under a wide array of circumstances. Thus, some of these concerns are, strictly speaking, well founded, but they do not amount to serious practical problems for social scientists as often as one might have supposed.

Finally, chapter 6 provides a summary of our advice for researchers seeking to formulate, estimate, test, and present interactive hypotheses in empirical research.

2

INTERACTIONS IN
SOCIAL SCIENCE

The interaction term received intense scrutiny, much of it critical, upon its introduction to social science. Althauser (1971) wrote, "It would appear, in short, that including multiplicative terms in regression models is not an appropriate way of assessing the presence of interaction among our independent variables" (466). Zedeck (1971) concurred, "The utility of the moderator variable research is limited by statistical problems, by the limited understanding of the statistical operation of moderators, and by lack of a rapid systematic approach to the identification of moderators" (307).

As Friedrich (1982) noted, early criticism of interactions focused on three concerns: difficulty in interpreting coefficients, colinearity among independent variables induced by the multiplication of terms, and the nature of measurement of independent variables (whether they be interval, ratio, or nominal scales). These concerns inspired some scholars (e.g., Althauser 1971; Zedeck 1971) to object to any usage of interactive terms. Others suggested alternative methods to incorporate interactions in models by rescaling variables to reduce colinearity (Allison 1977; Cronbach 1987; Dunlap and Kemery 1987; Smith and Sasaki 1979; Tate 1984).

Two and a half decades after the seminal article by Friedrich (1982) defending interactions, full and accurate understanding of the modeling, interpretation, and presentation of interactive hypotheses still eludes social scientists, even though multiplicative terms appear frequently in empirical analysis. For example, in a count of journal articles that appeared from

1996 to 2001 in the three top political-science journals,[1] we have found that 54 percent of articles use some statistical methods (defined as articles reporting any standard errors or hypothesis tests). Of these articles, 24 percent employ interactive terms. This amounts to about one-eighth of all articles published during this time.[2] Despite this appreciable and increasing use of interaction terms in empirical analysis, careful consideration of important classes of theoretical arguments in political science strongly suggests that they nonetheless remain considerably underutilized. Further, when interactions are employed in empirical work, several misunderstandings regarding their interpretation still permeate the field.

This widespread and perhaps expanding usage of interactions notwithstanding, we contend that still more empirical work should contain interactions than currently does, given the substance of many political-science arguments. Indeed, interactive arguments arise commonly in every empirical subfield in the social sciences. For political scientists, for example, interactive arguments appeal to scholars who study political institutions, to scholars who study political behavior, and perhaps especially to those who study the impact of institutions on political behavior, not to mention political economy, political culture, and all the other substantive areas of study within political science. These interactive arguments arise commonly in other disciplines: sociologists interested in the interactions between individuals and their social contexts, microeconomists examining the effect of public policies such as the minimum wage on different types of workers, macroeconomists studying the impact of fiscal or monetary policy under varying institutional conditions, and psychologists seeking to identify heterogeneity in individuals' reactions to experimental treatments. Interactions enable testing of these conditional-effect propositions.

In political science, for example, the core of most institutional arguments, reflecting perhaps the dominant approach to modern, *positive*[3] political science, implies interactive effects. In one influential statement of the approach, Hall (1986) states:

---

1. *American Political Science Review, American Journal of Political Science,* and *Journal of Politics.*

2. Incidentally, these shares likely dramatically understate the mathematical and technical nature of the field since our denominator includes pure-theory articles, formal and philosophical, and our numerator excludes formal theory. The share of statistical and formal-theoretical articles in these journals likely approaches 75 percent of all non-political-philosophy articles.

3. We use the term *positive* as opposed to *normative* here and do not intend it to connote *formal* necessarily.

the institutional analysis of politics . . . emphasizes institutional relationships, both formal and conventional, that bind the components of the state together and structure its relations with society . . . [I]nstitutions . . . refers to the formal rules, compliance procedures, and standard operating practices that structure the relationship between individuals in various units of the polity and economy . . . Institutional factors play two fundamental roles . . . [They] affect the degree of power that any one set of actors has over policy outcomes [. . . and they . . .] influence an actor's definition of his own interests, by establishing his . . . responsibilities and relationship to other actors . . . *With an institutionalist model we can see policy as more than the sum of countervailing pressure from social groups. That pressure is mediated by an organizational [i.e., institutional] dynamic.* (19; emphasis added)

Thus, in this approach, and inherently in all institutional approaches, institutions are *interactive* variables that funnel, moderate, or otherwise shape the political processes that translate the societal structure of interests into effective political pressures, those pressures into public-policy-making responses, and/or those policies into outcomes. Across all the methodological and substantive domains of institutional analysis, further examples abound:

[political struggles] *are mediated* by the institutional setting in which they take place. (Ikenberry 1988, 222–23; emphasis added)

[1] institutions *constrain* and *refract* politics but . . . are never the sole "cause" of outcomes. Institutional analyses do not deny the broad political forces that animate [. . . class or pluralist conflict, but stress how . . .] institutions *structure* these battles and, in so doing, influence their outcomes. [2. They] focus on how [the effects of] macrostructures such as class are *magnified* or *mitigated* by intermediate-level institutions . . . [they] help us integrate an understanding of general patterns of political history with an explanation of the *contingent* nature of political and economic development . . . [3] Institutions may be resistant to change, but *their impact on political outcomes can change* over time in subtle ways in response to shifts in the broader socioeconomic or political *context*. (Steinmo and Thelen 1992, 3, 11–12, 18; emphasis added)

the idea of structure-induced equilibrium is clearly a move [to-ward] incorporating institutional features into rational-choice ap-proaches. Structure and procedure *combine* with preferences to produce outcomes. (Shepsle 1989, 137; emphasis added)

Other recent examples include research that connects the societal structure of interests to effective political pressure through electoral in-stitutions: most broadly, plurality-majority versus proportional represen-tation (e.g., Cox 1997; Lijphart 1994; Ordeshook and Shvetsova 1994); research that studies how governmental institutions, especially those that affect the number and polarization of key policymakers (veto actors), shape policy-making responses to such pressures (e.g., Tsebelis 2002); research that stresses how the institutional configuration of the economy, such as the coordination of wage-price bargaining, shapes the effect of certain policies, such as monetary policy (see Franzese, 2003b, for a re-view). Examples could easily proliferate yet further.

In every case, and at each step of the analysis, from interest structure to outcomes (and back), the role of institutions is to *shape, structure, or condition*[4] the effect of some other variable(s)[5] on the dependent variable of interest. That is, most (probably all) institutional arguments are inher-ently interactive. Yet, with relatively rare exceptions, empirical evalua-tions of institutional arguments have neglected this interactivity in their models.

A more generic example further illustrates the common failure of em-pirical models to reflect the interactions that theoretical models imply. Po-litical scientists and economists consider principal-agent (i.e., delegation) situations interesting, problematic, and worthy of study because, if each had full control, agents would determine policy, y_1, by responding to some (set of) factor(s), X, according to some function, $y_1 = f(X)$. Principals, however, would respond to some different (set of) factor(s), Z, according to some function, $y_2 = g(Z)$. For example, the principal might be a current government, which responds to various political-economic conditions in setting inflation policy, and the agent an unresponsive central bank, as in Franzese (1999). Scholars then offer some arguments about how institu-

---

4. Extending the list of synonyms might prove a useful means of identifying interactive arguments. When one says *x* alters, *changes, modifies, magnifies, augments, increases, in-tensifies, inflates, moderates, dampens, diminishes, reduces, deflates,* and so on, some **effect** (of *z*) on *y*, one has offered an interactive argument.

5. Institutions seem most often to condition the impact of structural variables: for ex-ample, interest, demographic, economic, party-system structure, and so on. We suspect that this reflects some as-yet unstated general principle of institutional analysis.

tional and other environmental conditions determine the monitoring, enforcement, and other costs, C, principals must incur to force agents to enact $g(Z)$ instead of $f(X)$. In such situations, the connection between the realized policy, y, and the agent's preferred policy function, y_1, will depend on C or some function of C, say, $k(C)$. Similarly, the effect of the principal's policy function, y_2, on the realized policy will depend on C or some function of C, say, $[1 - k(C)]$. This reasoning suggests that the realized policy should be modeled as $y = k(C) \times f(X) + [1 - k(C)] \times g(Z)$ with $0 \leq k(C) \leq 1$ and $k(C)$ weakly increasing (see, e.g., Lohmann 1992, on the banks, governments, and inflation example). Thus, the effect on y of each $c \in C$ generally depends on X and Z, and the effect of each $x \in X$ and of each $z \in Z$ generally depends on C. That is, all factors that contribute to monitoring and enforcement costs modify the effect on y of all factors to which the principals and agents would respond differently, and, vice versa, the effect of all factors that determine monitoring and enforcement costs depends on all factors to which principals and agents would respond differently.[6] Most empirical models of principal-agent situations do not reflect this inherent interactivity.

For those who study individual or mass political behavior, opportunities to specify interactive hypotheses also abound. Scholars who argue that the effects of some set of individual characteristics (e.g., partisanship, core values, or ideology) depend on another set of individual characteristics (e.g., race, ethnicity, or gender) are proposing hypotheses that can and should be analyzed with interactive terms. Research questions that ask how the impact of some experimental treatment or environmental context (e.g., campaign or media communications) depends on the level of some individual characteristic (e.g., political awareness) likewise imply interactive hypotheses. Questions that explore how context (e.g., minority neighborhood composition or news media coverage of an issue) conditions the effect of some other predictor (e.g., racism) also reflect interactive hypotheses. Generally speaking, research questions that propose heterogeneity in how different types of individuals (or different microlevel units, even more generally) respond to their environments and institutional (i.e., macrolevel) contexts can and should be modeled interactively.[7]

6. Franzese (1999, 2002) shows how to use nonlinear regression to mitigate the estimation demands of such highly interactive propositions.

7. These last will also often imply *spatial interdependence;* see the following for methodological issues implied: Franzese and Hays (2005), Beck, Gleditsch, and Beardsley (2006), and contributions to *Political Analysis* 10(3). For multilevel contextual models, see the section "Random-Effects Models and Hierarchical Models" in Chapter 5, and the contributions to *Political Analysis* 13(4).

Interaction terms are widely used in statistical research in social science, and, in many more cases, theories suggest that interactions should be used although they are not. Despite their proliferation, some confusion persists regarding how to interpret these terms. Accordingly, we now provide practical advice to assist students and scholars to minimize this confusion.

3

THEORY TO PRACTICE

In this chapter, we provide guidance for constructing statistical models that map onto substantive theory. We discuss the implementation of statistical analyses to test the theory, and we provide advice on interpreting empirical results.

Specifying Empirical Models to Reflect Interactive Hypotheses

Theory should guide empirical specification and analysis. Thus, for instance, empirical models of principal-agent and other shared-policy-control situations should reflect the convex-combinatorial form, with its multiple implied interactions, as described earlier. Theoretical models of behavior that suggest that institutional or environmental contexts shape the effect of individual characteristics on behaviors and attitudes should likewise specify empirical models that reflect the hypothesized context conditionality in interactions.

To facilitate discussion, we will provide empirical examples from a variety of substantive venues. Our first empirical example comes from Gary Cox's *Making Votes Count* (1997). (More examples from other substantive venues and illustrating interactions of other types of variables will appear later.) Cox's justifiably acclaimed book makes several institutional arguments in which some political outcome, y, say, the effective number of parties elected to a legislature or the effective number of presidential candidates, is a function of some structural condition, x, say, the number of societal groups created by the pattern of social cleavages (e.g., the effective number of ethnic groups), and some institutional condition, z, say, the proportionality or district magnitude of the electoral system or the presence or absence of a presidential runoff system. Theory in this case very clearly implies that the relationship between y and x should be conditional upon z and, conversely, that the relationship between y and z should be conditional upon x. As Cox (1997) theorizes, for example, "A polity will have many parties only if it *both* has many cleavages *and* has a permissive enough electoral system to allow political entrepreneurs to base separate parties on these cleavages. Or, to turn the formulation around, a polity can have few parties either because it has no need for many (few cleavages) or poor opportunities to create many (a constraining electoral system)" (206). (See, Amorim Neto and Cox 1997; Cox 1997; Ordeshook and Shvetsova 1994 for empirical implementation.)

The standard linear-interactive model can reflect the propositions that x and z affect y and that the effects of x and of z on y each depend on the other variable. One simple way to write this (compound) proposition into a linear-regression model is to begin with a standard linear-additive model expressing a relation from x and z to y, along with an intercept, and then to allow the intercept and the coefficients on x and z each to depend on the level of z and x:[1]

$$y = \beta_0 + \beta_1 x + \beta_2 z + \varepsilon \tag{1}$$

$$\beta_0 = \gamma_0 + \gamma_1 x + \gamma_2 z$$

$$\beta_1 = \delta_1 + \delta_2 z$$

$$\beta_2 = \delta_3 + \delta_4 x$$

Combining these equations, we may express the model of y for estimation by linear regression in the standard linear-interactive manner:

$$y = \gamma_0 + \beta_x x + \beta_z z + \beta_{xz} xz + \varepsilon \tag{2}$$

As originally expressed in (1), the coefficients in this linear-interactive model (2) are $\beta_x = \gamma_1 + \delta_1$, $\beta_z = \gamma_2 + \delta_3$, $\beta_{xz} = \delta_2 + \delta_4$. More important, in this model, the effects of x and z on y depend on z and x, respectively, as an interactive theory would suggest.

Theory or substance might suggest several alternative routes to this same general model. For example, suppose we were to specify a system

---

1. We begin with the simplest case, where the effects of x and of z are deterministically dependent on, respectively, z and x. Subsequently, we relax this assumption to discuss probabilistic dependence (i.e., with error).

of relationships in which the effect of x on y and the intercept (conditional mean of y) depend on z:

$$y = \beta_0 + \beta_1 x + \varepsilon \qquad (3)$$

$$\beta_0 = \gamma_0 + \gamma_1 z$$

$$\beta_1 = \delta_1 + \delta_2 z$$

This is a common starting point for "multilevel" models in which some individual (microlevel) characteristic, x, is thought to produce microlevel outcomes or behaviors, y, although the mean of that outcome or behavior, β_0, and the effect, β_1, of that individual characteristic, x, may vary across contexts, which are described by a macrolevel characteristic, z. Combining these equations, we may express the following model for y:

$$y = \gamma_0 + \beta_x x + \beta_z z + \beta_{xz} xz + \varepsilon \qquad (4)$$

where $\beta_x = \delta_1$, $\beta_z = \gamma_1$, $\beta_{xz} = \delta_2$.

Note that the models actually estimated in (2) and (4) are identical, even though the theoretical/substantive stories told to derive the models from (1) and (3) seem to differ.[2] Each of these seemingly different theoretical stories yields the same mathematical model: the linear-interactive model (2).[3] This result demonstrates that, although the substance may determine which of these arguments is most intuitive to express, the distinction cannot be drawn mathematically. This mathematical ambiguity arises because the propositions being modeled are logically symmetric; that is, these statements all logically imply each other, and, in that sense, *they are identical; they cannot be distinguished because they are not distinct.* As Fisher (1988) writes, "prior theoretical specification is needed to interpret [in this sense] regression equations with product terms" (106). We concur but stress that the interpretive issues here are presentational and semantic because the alternatives are logical equivalents. These alternative theoretical stories may sound different in some substantive contexts, and some versions may seem more intuitive to grasp in certain contexts and others in other contexts. However, they are not actually alternatives; they are all the same tale.

---

2. To complete the list: a model in which y is a linear-additive function of z and the effect of z and the intercept depends on x, or one where the effect of x depends on z or the effect of z depends on x (and each effect may be nonzero when the other variable equals zero), also produces this same linear-interactive regression model.

3. Note: the linear-interactive model is not the only model form that would imply that the effects of x depend on z and vice versa, but, absent further theoretical elaboration that might suggest a more specific form of interaction, additive linear-interactive models like (2) are the logical, simple default in the literature.

Alternatively, one could propose a substantive argument that the effect of x on y depends on z but that z matters for y only insofar as it alters the impact of x and, in particular, z has no effect when x is equal to zero (not present). This *is* a change in the theoretical account of the relationship between the variables; it is a logically distinct argument, and it produces a truly different equation to be estimated:

$$y = \beta_0 + \beta_1 x + \varepsilon$$

$$\beta_1 = \delta_1 + \delta_2 z$$

$$y = \beta_0 + \beta_x x + \beta_{xz} xz + \varepsilon \tag{5}$$

where $\beta_x = \delta_1$, $\beta_{xz} = \delta_2$.

In this system of equations, we again see that z conditions the effect that x has on y and vice versa. However, by theoretical claim and ensuing model construction, z's sole effect is to determine the effect of x on y, and, in particular, movements in z have no effect on y when x is zero.[4] Scholars will typically think of z in this scenario as the *intervening variable:* intervening in x's relationship to y. However, notice that just as a value of x exists, namely, $x = 0$, where the effect of z is zero, a value of z exists, namely, $z = -\beta_x/\beta_{xz}$, where the effect of x is zero. The substance of the context at hand may suggest whether to conceive $x = 0$ or $z = -\beta_x/\beta_{xz}$, or, for that matter, some other value of x or z, as the base from which to decide whether x or z is the one *intervening* in the other's relationship with y. Mathematically that determination is once again arbitrary because, *logically, all interactions are symmetric.*[5] Given this logical symmetry, x and z must necessarily *both* intervene in the other's relationship to y. In this sense, the language of one variable being the intervening or moderating variable and the other being the one moderated may be best avoided; if an interaction exists, then all variables involved intervene or moderate in the others' relations to y.

The preceding equations assume that the effect of x on y depends on z and the effect of z on y depends on x *deterministically,* that is, without error. This might seem odd, given that our model proposes that x and z predict y only with error (hence the inclusion of the term ε), but the subsequent equations propose that the effect of x on y and of z on y each

4. We discuss this type of model further in the first section of chapter 4.

5. Mathematically, the proof of this logically necessary symmetry in all interactions is simply

$$\frac{\partial\left(\frac{\partial f(x,z)}{\partial x}\right)}{\partial z} \equiv \frac{\partial^2 f(x,z)}{\partial x \partial z} \equiv \frac{\partial^2 f(x,z)}{\partial z \partial x} \equiv \frac{\partial\left(\frac{\partial f(x,z)}{\partial z}\right)}{\partial x} \quad \forall\, f(x,z).$$

depend on the other variable *without error*. We can easily amend the linear-interactive model to allow a more logically consistent stochastic conditioning of the effects of variables by the others' levels thus:

$$y = \beta_0 + \beta_1 x + \beta_2 z + \varepsilon$$

$$\beta_0 = \gamma_0 + \gamma_1 x + \gamma_2 z + \varepsilon_0$$

$$\beta_1 = \delta_1 + \delta_2 z + \varepsilon_1$$

$$\beta_2 = \delta_3 + \delta_4 x + \varepsilon_2$$

Combining these equations allows expressing y for regression analysis in the now-familiar

$$y = \gamma_0 + \beta_x x + \beta_z z + \beta_{xz} xz + \varepsilon^* \tag{6}$$

where $\varepsilon^* = \varepsilon + \varepsilon_0 + \varepsilon_1 x + \varepsilon_2 z$, $\beta_x = \gamma_1 + \delta_1$, $\beta_z = \gamma_2 + \delta_3$, $\beta_{xz} = \delta_2 + \delta_4$.[6]

The composite residual ε^* in (6) retains zero expected value and noncovariance with the regressors x, z, and xz provided that its components, ε, ε_0, ε_1, and ε_2, do so. These key assumptions of the classical linear-regression model (CLRM) ensure unbiasedness and consistency of ordinary least squares (OLS) coefficient estimates. However, this compound residual does not retain constant variance, since it waxes and wanes as a function of x and z. This heteroskedasticity undermines the efficiency of the OLS coefficient estimates and the accuracy of OLS standard errors. In other words, if the conditionalities of the x and z relationships with y themselves contain error, then the standard linear-interactive model has heteroskedastic error even if the individual stochastic terms comprising its compound residual are homoskedastic. Thus, OLS coefficient estimates are unbiased and consistent but not efficient. The OLS standard-error estimates, on the other hand, are incorrect,[7] but, as we show later, these problems are often easy to redress satisfactorily with familiar techniques. We return to this technical concern in the section "Random-Effects Models and Hierarchical Models" in chapter 5, because this concern often underlies calls for random-coefficient or linear-hierarchical

6. Note that the terms involving $\varepsilon_1 x$ and $\varepsilon_2 z$ can be removed from the expression for the composite error, ε^*, and replaced by appending $+\varepsilon_1$ to the expression for β_x and $+\varepsilon_2$ to that for β_z, to give another common expression of the random-coefficients/random-effects model.

7. To be precise, OLS standard-error estimates, as estimates of the true variation across repeated samples of the OLS coefficient estimates under the CLRM assumptions, are always inefficient in the presence of any heteroskedasticity, and, when the heteroskedasticity is a function of the regressors, as is the case here, they are biased and inconsistent as well.

models. For now, we proceed assuming the researcher estimates a model like (4) by OLS.

Let us return to our example of electoral systems, social cleavages, and the number of parties or candidates to illustrate the preceding discussion. We follow Cox's analysis of the effects of presidential-runoff systems (*Runoff*) and the effective number of ethnic groups in a society (*Groups*) on the effective number of presidential candidates (*Candidates*) that emerges in a presidential democracy.[8] The theory suggests that the impact of social cleavages on the effective number of candidates depends on whether a runoff system is used and, symmetrically, that the impact of the runoff system on the effective number of candidates depends on the number of societal groups. (Recall that these are logically two sides of the same proposition.) The confluence of a high number of social cleavages and a runoff system is hypothesized to produce a high effective number of presidential candidates, because the number of societal groups increases the potential number of parties and the runoff system attenuates the incentives for preelection coalition building between such groups. Given this theoretical structure, we can specify the following model for estimation:

$$Candidates = \beta_0 + \beta_G Groups + \beta_R Runoff + \beta_{GR} Groups$$
$$\times \; Runoff + \varepsilon \qquad (7)$$

The data set includes information from sixteen presidential democracies in 1985.[9] The dependent variable, *Candidates*, the effective number of presidential candidates, ranges from 1.958 to 5.689, with a mean of 3.156 and a standard deviation of 1.202. The independent variable, *Groups*, the effective number of ethnic groups in a society,[10] ranges from 1 to 2.756, with a mean of 1.578 and a standard deviation of 0.630. The independent variable, *Runoff*, indicates the presence or absence of a runoff system for the presidential election; this dummy variable takes the value of zero if the system does not employ runoffs and one if it does use

---

8. Effective numbers are simply size-weighted counts of items. The effective number of social groups, for example, is $\left(\sum_{i=1}^{n} g_i^2\right)^{-1}$, where g_i is the group i's fraction of the population. The effective number of candidates is $\left(\sum_{i=1}^{n} v_i^2\right)^{-1}$, where v_i is candidate i's fraction of the vote total.

9. We selected this data set because it is freely available (at http://dodgson.ucsd .edu/lij/pubs/) so researchers can easily replicate our results and because of its very manageable size. The small N, however, makes finding any strong statistical significance rather unlikely, but weak significance hardly hampers our pedagogical purposes.

10. To avoid some tiresome repetition, we henceforth drop the adjectives *effective*, although they remain applicable.

them. In this sample of sixteen presidential democracies, exactly half have a runoff system. The OLS regression results appear in table 1.

How do we interpret these results? What do these estimated coefficients mean? The next section provides guidance on these questions.

Interpreting Coefficients from Interactive Models

In the simple linear-additive regression, $y = \beta_0 + \beta_x x + \beta_z z + \varepsilon$, the effect of the variable, x, on y is simply its coefficient, β_x. When x rises by one unit, ceteris paribus, y rises by β_x. Likewise for z, its effect on y is its coefficient, β_z. In this case—and *only* in the purely linear-additive regression case—the *coefficient* on a variable and the *effect* on the dependent variable of a unit increase in that independent variable (ceteris paribus and controlling for other regressors) are identical.

In interactive models, as in all models beyond the strictly linear-additive, this equivalence of coefficient and effect no longer holds. In an attempt to cope with this change, current practice in interpreting interactive effects often substitutes some vague and potentially misleading terms, such as *main effects* and *interactive effect, direct effects* and *indirect effect,* and *independent effects* and *total effect,* for the coefficients on

TABLE 1.   OLS Regression Results, *Number of Presidential Candidates*

	Coefficient (standard error) p-Value
Ethnic Groups	−0.979 (0.770) 0.228
Runoff	−2.491 (1.561) 0.136
Ethnic Groups × Runoff	2.005 (0.941) 0.054
Intercept	4.303 (1.229) 0.004
N (degrees of freedom)	16 (12)
Adjusted R^2	0.203
$P > F$	0.132

Note: Cell entries are the estimated coefficient, with standard error in parentheses, and two-sided p-level (probability $|T| > t$) referring to the null hypothesis that $\beta = 0$ in italics.

x and z in the first case and on xz in the second. Such terminology is usually unhelpful at best, misleading or incorrect at worst.[11]

Instead, we encourage researchers to recall that each variable involved in the interaction terms of interactive models has multiple effects, not any single, constant effect, such as might be given somehow by a single coefficient, nor a *main* effect and an *interactive* effect, such as might be given by some pair of coefficients, but multiple, different effects depending on the levels of the other variable(s) with which it interacts. When a researcher argues that the effect of some variable x on y depends on z, he or she is arguing that x has different effects on y, depending on the specific values of z. In the interactive case, the effects of x on y are therefore not any single constant, like the coefficient β_x on x in the simple linear-additive model. The effects of x on y vary. They depend on the coefficients on x and xz, as well as the value of z. To restate the general principle: outside of the purely linear-additive model, *coefficients are not effects*. The effect of x on y, as we elaborate subsequently, is the derivative, $\partial y / \partial x$, or the difference/change, $\Delta y / \Delta x$, which will only equal the coefficient on x by itself in the purely linear-additive case.

Terming one coefficient the *main* effect and another the *interactive* effect thus perilously confuses coefficients for effects. Substantively, there may in fact be nothing whatsoever "main" or "direct" about the particular effect to which the coefficient on x actually does refer. Researchers cannot appropriately refer to the coefficient on x as "the main effect of x" or "the effect of x . . . independent of z" or "considered independently of z" or, certainly not, "controlling for z." The coefficient on x is just one effect x may have, namely, the effect of x at $z = 0$. That is, the coefficient on x gives the estimated effect of a unit change in x, *holding z fixed at zero*. We note that this zero value of z may have nothing at all "main" about it. It may fall outside the range of what appears in the sample, or it could even be logically impossible! The effect of x on y at $z = 0$ is obviously not "independent of z"; in fact, it is connected with a particular value of z. This effect of x on y when $z = 0$ is also a very different thing from the effect of x on y "controlling for z." The simple linear-additive multiple-regression model estimates a single, constant "effect of x on y, controlling for z." The linear-interactive model estimates the effect of x on y as a function of z.

Our empirical example illustrates and clarifies these points. The esti-

---

11. Note that some of this terminology also refers to path-analytic models, which specify that some variable x affects the level (rather than, or in addition to, the effect) of some variable z that then determines y. This overlap in terminology provides even more confusion for the researcher.

mated coefficient on *Runoff* ($\hat{\beta}_R = -2.491$) gives the estimated effect of runoff elections on the number of presidential candidates for the specific case where *Groups* takes a value of zero. But the number of societal groups never takes the value of zero in the sample; in fact, the number of ethnic groups in a society cannot logically equal zero. Thus, an interpretation of $\hat{\beta}_R$, the estimated coefficient on *Runoff*, as the "main" effect of a runoff system is nonsensical; far from a "main" effect, this is actually the effect at a value of ethnic heterogeneity that does not, and indeed could not, exist.

If, however, *Groups* were rescaled to include a value of zero, for example, by subtracting some constant value, such as the mean, and calling the resulting variable *Groups**, then the estimated coefficient $\hat{\beta}_{R*}$ would be the estimated effect of *Runoff* when the rescaled variable *Groups** takes the value of zero. This is assuredly logically possible and in sample now, but the notion that the effect at this particular substantive value of ethnic heterogeneity is somehow "main" would remain strained and potentially misleading. That the effect of some variable when its moderating variable happens to be at its mean should be called a "main effect" while all the other effects at all the other logically permissible or empirically existent values are something other than "main" seems an unnecessary and possibly misleading substantive imposition, especially since the theoretical and substantive point of the interaction model in the first place is that the effects of the interacting variables vary depending on each other's values. We return to this topic of mean-rescaling interactive variables in the first section of chapter 4.

Symmetrically, the estimated coefficient $\hat{\beta}_G$, the coefficient on *Groups*, refers to our estimate of the effect of the number of ethnic groups when *Runoff* equals zero. This value does logically and empirically exist, and so the estimated value of $\hat{\beta}_G = -0.979$ tells us something substantively relevant. It reports an estimate that, in a system without runoffs, the number of ethnic groups has a negative impact on the number of presidential candidates. Specifically, an increase of 1 in the number of ethnic groups is empirically associated with a 0.979 reduction in the number of presidential candidates, *in systems without runoff elections*. (We find this result substantively puzzling, but that is the estimate.) Note, however, that the coefficient $\hat{\beta}_G$ only tells part of the story—it only reveals the estimated effect of *Groups* in one condition: when *Runoff* equals zero.

The researcher who equates a coefficient in an interactive model to an effect is thus treading on hazardous ground. At best, the researcher will be telling a story about an effect that applies to only one of several possible conditions (e.g., when $z = 0$ or when $z = \bar{z}$). At worst, the researcher will be telling a story about an effect that applies in no logically possible condition—an effect that is logically meaningless. In short, put simply, and re-iterating this crucial point: outside the simplest purely linear-additive case, *coefficients* and *effects* are different things.

We suggest two effective and appropriate methods of interpreting results from interactive models: differentiation (which requires working knowledge of entry-level calculus) and differences in predicted values (which does not).

Interpreting Effects through Differentiation

Consider the following standard linear-interactive regression model:

$$y = \gamma_0 + \beta_x x + \beta_z z + \beta_{xz} xz + \varepsilon \tag{8}$$

The effects of an independent variable, x, on the dependent variable, y, can be calculated by taking the first derivative of y with respect to x (as suggested by, e.g., Friedrich 1982; Stolzenberg 1979). This is a direct and simple means of identifying the effects of x on y or the effects of z on y because first derivatives or first differences, $\partial y/\partial x$ and $\partial y/\partial z$, or $\Delta y/\Delta x$ and $\Delta y/\Delta z$, *are* effects. One may, in fact, read $\partial y/\partial x$ (or $\Delta y/\Delta x$), for example, as "the change in y, ∂y (or Δy), induced by a marginal (derivative) or unit (difference) increase in x, ∂x (or Δx), all else held constant." Differentiation is a simple, reliable, methodical way of calculating inter-active effects. To help it fulfill its promise of simplicity and to reduce the tendency to induce mistakes, we provide a table of basic differentiation rules in appendix A.

In the standard linear-interactive model (8), the first derivatives of y with respect to x and z are

$$\partial y/\partial x = \beta_x + \beta_{xz} z \tag{9}$$

$$\partial y/\partial z = \beta_z + \beta_{xz} x \tag{10}$$

As (9) and (10) exemplify, the first derivative of (8) with respect to x and z yields the conditional effect of those variables directly. Derivatives *are* effects, whether in basic linear-additive regression models, when they yield just the coefficient on the variable of interest, or in linear-interac-tive models like (8), when they give expressions like (9) and (10) involv-ing two coefficients and the other interacting variable. This generalizes to any other model regardless of its functional form.

The effect of x on y in an interactive model like (8) is $\beta_x + \beta_{xz}z$, which reflects the conditional argument underlying that model. As noted earlier, β_x is merely the effect of x on y when z happens to equal zero, and it is neither necessarily the "main" effect in any sense nor the effect "independent of" or "controlling for" z. Nor, we now add, does β_{xz} embody the "interactive" effect of x or of z exactly, as often suggested. The coefficient β_{xz} indicates by how much the effect of x on y changes per unit increase in z. It also indicates the logically and mathematically identical amount by which a unit increase in x changes the effect of z on y. Neither is precisely an effect. They are statements of how an effect *changes:* that is, an effect on an effect. The sign and magnitude of β_{xz} tell us how the effect of x on y varies according to values of z. In an interactive model, indeed in any model, the effect of a variable, x, on y is $\partial y / \partial x$. Here that effect is $\beta_x + \beta_{xz}z$. One cannot distinguish some part of this conditional effect as main and another part as interactive.

Returning to our empirical example of the interaction between institutional structure and social cleavages in determining the number of presidential candidates, we are now prepared to interpret the results using differentiation. Recall the results from our OLS regression:[12]

$$\widehat{Candidates} = 4.303 - 0.979(Groups) - 2.491(Runoff)$$

$$+ 2.005(Groups \times Runoff) \tag{11}$$

Applying (9) and (10), we see that

$$\partial \hat{y}/\partial G = -0.979 + 2.005(Runoff) \tag{12}$$

$$\partial \hat{y}/\partial R = -2.491 + 2.005(Groups) \tag{13}$$

Thus, the effect of societal groups on the number of presidential candidates varies with the presence or absence of a runoff, and the effect of a runoff on the number of presidential candidates varies with the number of ethnic groups in society. These conditional effects can be easily calculated by inserting substantively relevant values for the variables of interest into equations (12) and (13).

Recall that *Runoff* takes only two values: zero in the absence and one in the presence of a runoff system. Hence, we use (12) to recalculate the

---

12. Although, technically, one cannot strictly differentiate with respect to noncontinuous variables, such as dummy variables, one can proceed ignoring this technicality without being misled. (Do remember, however, that *marginal* increases cannot actually occur, only *unit* increases from zero to one can.) Alternatively, one can calculate differences in predicted values, which we discuss next. For more detail, see note 17.

conditional effect of *Groups* on the number of candidates for these two substantively relevant values of *Runoff*. When $Runoff = 0$, $\partial \hat{y}/\partial G = -0.979 + 2.005 \times 0 = -0.979$. When $Runoff = 1$, $\partial \hat{y}/\partial G = -0.979 + 2.005 \times 1 = 1.026$. In the absence of a runoff, the estimated effect of ethnic groups is negative; in the presence of a runoff, the estimated effect of ethnic groups is positive. (The "Linking Statistical Tests with Interactive Hypotheses" section of this chapter discusses the standard errors and statistical significance of these estimated effects, which, like the effects themselves, vary with the level of the conditioning variable.)

Symmetrically, we can calculate the conditional effect of *Runoff* on the number of presidential candidates by inserting substantively relevant values of *Groups* into (13). Recall that *Groups* ranges from 1 to 2.756 in our data set. We should present the estimated effects of *Runoff* over a substantively revealing set of values for *Groups*: for example, over the sample range of values of *Groups*; or at evenly spaced intervals starting from the sample minimum to some substantively meaningful and revealing high value; or at the minimum, mean, and maximum; or at the mean, the mean plus and the mean minus a standard deviation or two.

To take one of these options, we calculate conditional effects when *Groups* ranges from 1 to 3, at evenly spaced intervals of 0.5, which yields the following estimated conditional effects:[13]

When Groups = 1: $\partial \hat{y}/\partial R = -2.491 + 2.005 \times 1 = -0.486$

When Groups = 1.5: $\partial \hat{y}/\partial R = -2.491 + 2.005 \times 1.5 = 0.517$

When Groups = 2: $\partial \hat{y}/\partial R = -2.491 + 2.005 \times 2 = 1.520$

When Groups = 2.5: $\partial \hat{y}/\partial R = -2.491 + 2.005 \times 2.5 = 2.522$

When Groups = 3: $\partial \hat{y}/\partial R = -2.491 + 2.005 \times 3 = 3.525$

At the sample minimum (when the society has only one ethnic group), a runoff system has a negative effect on the number of presidential candidates (which, again, seems substantively odd), but as the number of ethnic groups rises, the runoff begins to affect the number of presidential candidates positively (which is more sensible). The size of the effect grows as ethnic groups become more numerous (also sensible). Again, the standard errors of these estimated effects and whether the effects are statistically significant are matters we will discuss subsequently.

---

13. Although the sample maximum is 2.756, *Ethnic Groups* does extend beyond this value in some of the nonpresidential systems that Cox (1997) analyzes.

Interpreting Effects through Differences in Predicted Values

A second strategy for examining the effects of x and z on y consists of examining differences in predicted values of y for logically relevant and substantively meaningful values of x and z. This strategy does not require the researcher to have any knowledge of calculus; it is a bit more tedious but quite serviceable in its own right. Predicted values of y, denoted as $\hat{y}$, can be calculated by substituting the estimated values for the coefficients along with logically relevant and substantively revealing values of the covariates of interest into the theoretical model (equation (8)) and substituting in estimated coefficient values:

$$\hat{y} = \hat{\gamma}_0 + \hat{\beta}_x x + \hat{\beta}_z z + \hat{\beta}_{xz} xz \tag{14}$$

We can now calculate $\hat{y}$ at varying values of x (between, say, x_a and x_c) while holding z constant at some meaningful value (e.g., its mean value or some other substantively relevant value; if z is a dummy, for example, zero and one are meaningful). By doing so, the researcher can calculate how changes in x (from x_a to x_c) cause changes in $\hat{y}$ (from $\hat{y}_a$ to $\hat{y}_c$). Recall that as x changes from x_a to x_c, while z is held at some meaningful value, say, z_0, this also implies that xz changes from $x_a z_0$ to $x_c z_0$. The predicted values, $\hat{y}_a$ and $\hat{y}_c$, can be calculated as follows:

$$\hat{y}_a = \hat{\gamma}_0 + \hat{\beta}_x x_a + \hat{\beta}_z z_0 + \hat{\beta}_{xz} x_a z_0 \quad \text{and}$$

$$\hat{y}_c = \hat{\gamma}_0 + \hat{\beta}_x x_c + \hat{\beta}_z z_0 + \hat{\beta}_{xz} x_c z_0$$

The change in predicted values can be calculated as the difference between $\hat{y}_a$ and $\hat{y}_c$:

$$\hat{y}_c - \hat{y}_a = \hat{\gamma}_0 + \hat{\beta}_x x_c + \hat{\beta}_z z_0 + \hat{\beta}_{xz} x_c z_0 - (\hat{\gamma}_0 + \hat{\beta}_x x_a + \hat{\beta}_z z_0$$

$$+ \hat{\beta}_{xz} x_a z_0) \; \hat{y}_c - \hat{y}_a = \hat{\beta}_x (x_c - x_a) + \hat{\beta}_{xz} z_0 (x_c - x_a) \tag{15}$$

Symmetrically, the researcher can identify how $\hat{y}$ moves with changes in z (and xz) when x is held at some meaningful value.

In our example, we can examine how the predicted number of presidential candidates changes as we increase the number of ethnic groups in the presence and in the absence of runoff elections:

$$\overline{Candidates} = 4.303 - 0.979(Groups) - 2.491(Runoff)$$

$$+ \; 2.005(Groups \times Runoff)$$

When $Groups = 1$ and $Runoff = 0$, we calculate the predicted number of candidates, $\hat{y}$ as

$$(\hat{y} \mid Groups = 1, Runoff = 0) = 4.303 - 0.979 \times 1 - 2.491 \times 0$$

$$+ \; 2.005 \times 1 \times 0 = 3.324$$

TABLE 2.   Predicted *Number of Presidential Candidates*

	Runoff = 0	*Runoff* = 1
Groups = 1	3.324	2.838
Groups = 1.5	2.835	3.351
Groups = 2	2.345	3.865
Groups = 2.5	1.855	4.378
Groups = 3	1.366	4.891

Table 2 presents the predicted number of candidates, as *Groups* ranges from one to three, when *Runoff* takes values of zero and one.

From such a table, the researcher can discern how the independent variables, *Groups* and *Runoff*, influence the predicted dependent variable, *Candidates*. By looking across single rows, we see the effect of the presence of a runoff system at the given number of *Groups* associated with each row. In the first row, for example, when the value of *Groups* is at its minimum (one), a runoff system has a small and negative effect, decreasing the number of parties by −0.486 (that same substantively odd result again). When the value of *Groups* is at a higher value, say, 2.5 (row 4), the impact of a runoff system is larger in magnitude and positive: in a polity with 2.5 social groups, a runoff system is estimated to increase the number of presidential candidates by (a substantively sensible) 2.523.

By looking down single columns, we see the effects of changes in the number of ethnic groups in the absence or in the presence of a runoff system. In the absence of a runoff system, a rise in the number of ethnic groups from, say, one to three coincides (oddly) with a decline in the number of presidential candidates from 3.324 to 1.366. In the presence of a runoff system, however, a rise in the number of ethnic groups from, say, one to three coincides (sensibly) with an increase in the number of presidential candidates (from 2.838 to 4.891). Subsequently, we address standard errors for these estimated changes and whether they are statistically distinguishable from zero.

Interpreting Interactive Effects Involving Different Types of Variables

Our advice on interpretation applies generally across essentially[14] all types of variables that scholars might analyze—dummy variables, discrete

---

14. Ordinal independent variables mildly complicate interpretation of linear-regression estimates, whether of purely linear-additive or of linear-interactive form, because linear-regression treats all independent-variable information as cardinal. In practice, researchers often assume ordinal variables to give cardinal, or close enough to cardinal,

variables, continuous variables, and so on—and any nonlinear transformations (such as $\ln(x)$ or x^2, or $\{v = 1$ if $x < x_0$, $v = 0$ if $x \geq x_0\}$). Furthermore, virtually all permutations of interactions between these types of variables are logically and empirically possible,[15] and all can be interpreted using one or both of our approaches. One need not learn different interpretational techniques for each different variable and interaction type; the preceding discussion fully suffices, as we illustrate next. (Optimal presentational efficacy will often suggest different graphical and/or tabular approaches for different applications as the next section suggests and illustrates.)

Our first empirical example illustrates one of these possible permutations: an interaction between a dummy variable (*Runoff*) and a continuous variable (*Groups*). To illustrate more of the rich range of possibilities, we now introduce some additional empirical cases.

Our second example derives from public-opinion research into partisan and gender gaps in support for social welfare (e.g., Box-Steffensmeier, De Boef, and Lin 2004; Shapiro and Mahajan 1986). Theory suggests that social-welfare attitudes derive from a set of individual-level characteristics, such as partisan orientation, ideology, gender, race, and income, and that the effect of one or more of these characteristics, such as partisanship, might depend on some other characteristic, such as gender. Partisanship is strongly related to support for social-welfare programs; for example, in the United States, Republicans are less supportive of these programs than Democrats. Gender is also strongly related to support for social-welfare programs, with females generally more supportive than males. However, if partisan and gender influences are complements or substitutes in opinion formation regarding social welfare, then the effect of partisanship among females will differ from that among males. Symmetrically, the effect of gender will differ among Republicans compared with the gender effect among Democrats. A

---

information. Nominal variables complicate linear-regression interpretation similarly. For binary nominal (i.e., dummy) variables, the researcher need only remember the variable's binary nature when considering substantively meaningful ranges of or changes in those variables. Since a unit change is the only change possible, whether that dummy offers nominal, ordinal, or cardinal information does not alter the mechanics of interpretation. For nominal variables with more than two categories, increases or decreases in a variable's value do not correspond to any substantive notion of increase or decrease, and so their direct use in linear regression, again whether of purely linear-additive or of linear-interactive form, is not even approximately appropriate. For use in regression analysis, researchers would first decompose such multinomial variables into sets of binary variables, each indicating one of the categories.

15. The one exception is that a dummy variable interacted with itself just gives itself back, and so x and x^2 are identical if x is a dummy.

standard linear-interactive model like (8) would enable a test of such theoretical propositions.

We analyze such a model using data from the 2004 American National Election Studies. The dependent variable is an additive index of support for the social-welfare state.[16] The independent variables are an indicator (dummy) for gender (one if female, zero if male), an indicator for partisanship (one if Republican, zero if Democrat; with all others excluded for ease of exposition), and the interaction of those two variables.

$$Opinion = \beta_0 + \beta_F Female + \beta_R Republican + \beta_{FR} Female$$
$$\times Republican + \varepsilon \qquad (16)$$

The OLS results appear in table 3. Note that this analysis features an interaction between two dummy variables. Differentiation (derivatives) will produce the correct expressions for the conditional effects, but calculating differences in predicted values might make more intuitive sense given the binary nature of the variables.[17] As such, these OLS results can be easily interpreted by comparing the predicted support for the social-welfare state for each of the four categories supplied by the multiplication of the two binary variables (male Democrat, female Democrat, male Republican, and female Republican).

The predicted values in table 4 suggest that there is little difference in the social-welfare support of male and female Democrats but that a gender gap does exist in support for social welfare between male and female Republicans. The gender gap is thus contingent upon partisanship. Con-

---

16. We provide this very simple example for pedagogical purposes; a more fully specified model would of course be more compelling. The dependent variable is compiled from support for services and spending; government provision of jobs and a standard of living; and support for federal spending on welfare programs, social security, public schools, child care, and assistance to the poor, rescaled to range from zero (least supportive) to one (most supportive).

17. Recall that derivatives are the limit of $\Delta y/\Delta x$ as Δx approaches zero. For a dichotomous variable, this is intuitively unappealing; given that the variable takes only two discrete values, 0 and 1, Δx can only be 1, 0, or -1. However, as Greene (2003) notes, "The computation of the derivatives of the conditional mean function [i.e., the regression equation] is useful when the variable in question is continuous and often produces a reasonable approximation for a dummy variable" (676). Indeed, the differentiation method will produce the correct mathematical formula for the conditional effects of a marginal change in x, and so the only issue here involves the meaningfulness of a marginal change. For linear interactions, one can simply determine the formula for the conditional effect by differentiation and then consider only discrete changes in the conditioning variable. For nonlinear interactions, however, the amount by which the conditional effect changes as the indicator or other discrete conditioning variable increases by one will not be constant over that unit range, and so the effect of a marginal change is not as substantively interesting and the difference method is more revealing.

versely, partisanship has a larger effect among men than women; male Republicans and male Democrats are farther apart than female Republicans and female Democrats. The degree of partisan polarization is contingent upon gender. Subsequently, we address the statistical uncertainty of these estimates.

Other interactions may involve the product of two continuous variables. Our third empirical example considers the duration of parliamentary governments and features this type of interaction. The dependent variable is the average duration of governments in the post–World War II era, in months, and takes values between 11 and 45.1. We model it as a function of the postwar average number of parties in government (*NP*), which ranges from 1 to 4.3; the postwar average parliamentary support for government in the legislature (i.e., the percentage of lower house

TABLE 3. OLS Regression Results, *Support for Social Welfare*

	Coefficient (standard error) p-Value
Female	−0.0031 (0.0144) *0.828*
Republican	−0.2205 (0.0155) *0.000*
Female × Republican	0.0837 (0.0214) *0.000*
Intercept	0.7451 (0.0110) *0.000*
N (df)	1,077 (1,073)
Adjusted R^2	0.223
$P > F$	0.000

Note: Cell entries are the estimated coefficient, with standard error in parentheses, and two-sided p-level (probability $|T| > t$) referring to the null hypothesis that $\beta = 0$ in italics.

TABLE 4. Predicted *Support for Social Welfare*

	Democrats (Republican = 0)	*Republicans (Republican = 1)*
Males (Female = 0)	0.745	0.525
Females (Female = 1)	0.742	0.605

seats held by the governing party or parties) (PS), which ranges from 41.1 to 80.4; and the level of party discipline (PD), an indicator for high levels of party discipline.[18] We specify an interaction between the number of parties in government and the parliamentary support for government, with the idea that as the degree of support for the governing party increases, the effect of the number of parties in government on the duration of government will likely decline (and vice versa). The term PD serves as a control; later we expand this model to illustrate other issues. Here, we estimate the following model:

$$\text{Government Duration} = \beta_0 + \beta_{np}NP + \beta_{ps}PS + \beta_{npps}NP \times PS$$
$$+ \beta_{pd}PD + \varepsilon \tag{17}$$

The OLS results appear in table 5. Both differentiation and differences in predicted values are useful in interpreting the results of an analysis featuring an interaction between two continuous variables. Differentiating, the effect of NP on the duration of governments is

$$\partial \hat{y}/\partial NP = \hat{\beta}_{np} + \hat{\beta}_{npps}PS = -31.370 + 0.468(PS)$$

The estimated coefficient $\hat{\beta}_{np} = -31.370$ suggests that the effect of the number of governing parties on government duration is -31.370 when $PS = 0$, but setting parliamentary support to zero is a substantively meaningless value, thus reinforcing our warning that coefficients are not the same as effects in the linear-interactive model. Increases in parliamentary support attenuate this negative effect of the number of parties on government duration (as hypothesized) until parliamentary support reaches a level of 67.02. At this point, the effect has reached zero: $\partial \hat{y}/\partial NP = 0$. When parliamentary support exceeds 67.02, the effect of the number of parties on government duration becomes positive. A positive effect seems substantively odd until we consider that only grand coalitions encompassing all or most of parliament would typically exceed such a high level of parliamentary support;[19] grand coalitions including more parties intuitively might indeed last longer than grand coalitions of fewer parties, which perhaps exclude some, thereby violating such coalitions' justifying principle. This example provides an interesting case where the effect of some variable x is negative in one range of z, crosses zero, and then becomes positive in another range of z. It also illustrates the importance of

18. The dummy variable PD reflects our own coding of party discipline in these democracies.

19. In the sample, Austrian and especially Swiss governments exceed 67 percent average parliamentary support appreciably, and governments in Luxembourg do so slightly. Reinforcing the explanation in the text, Swiss governments serve terms that are not determined by standard parliamentary processes.

considering conditional effects at substantively meaningful values in the sample.

Similarly, the effect of the degree of parliamentary support on government duration is

$$\partial\hat{y}/\partial PS = \hat{\beta}_{ps} + \hat{\beta}_{npps}NP = -0.586 + 0.468(NP)$$

The estimated effect of *PS* on government duration is negative when *NP* assumes a zero value (but this, too, is a substantively meaningless value in this example). At the meaningful minimum value of $NP = 1$, the estimated effect of *PS* is near zero, which is substantively sensible; single-party governments tend to last to term, regardless of their margin. The conditional effect of parliamentary support on government duration crosses zero when $NP = 1.25$ and is increasingly positive as *NP* increases further. Governments tend to endure longer as their parliamentary support increases, and this is especially so for multiparty governments, likely because governments of more parties are more easily fractured by events and circumstances and so have greater need of greater parliamentary support to survive the vicissitudes of coalition politics.

As table 6 exemplifies, these results can be interpreted equivalently by

TABLE 5. OLS Regression Results, *Government Duration*: Simple Linear-Interaction Model

	Coefficient (standard error) *p*-Value
Number of Parties (NP)	−31.370 (11.345) 0.013
Parliamentary Support (PS)	−0.586 (0.454) 0.214
Number of Parties × Parliamentary Support (NP × PS)	0.469 (0.186) 0.022
Party Discipline (PD)	9.847 (3.204) 0.007
Intercept	59.273 (26.455) 0.039
N (df)	22 (17)
Adjusted R^2	0.511
$P > F$	0.002

Note: Cell entries are the estimated coefficient, with standard error in parentheses, and two-sided *p*-level (probability $|T| > t$) referring to the null hypothesis that $\beta = 0$ in italics.

comparing the predicted government durations at varying meaningful levels of *NP* and *PS*, while holding any other variables in the model (*PD* in this case) also at substantively meaningful values (e.g., the sample mean or mode; in this case, we hold *PD* to a value of one).

Reading down the first column of calculated values, one sees a very modestly negative (i.e., near zero) estimated effect of *PS* under single-party government; across the entire forty-point sample range of *PS*, government duration declines by only 4.71 months (from 33.05 to 28.34). However, this effect intuitively reverses sign and grows substantially as the number of governing parties increases. When governments average three parties, predicted duration increases by a substantial 32.78 months as *PS* expands from its sample minimum 40 to maximum 80 percent. Reading across each of the rows, we see the estimated effects of the number of governing parties at given levels of parliamentary support. Intuitively, increases in *NP* are associated with declines in government duration over most values of governing support (although they are associated with longer government durations at the very high values of *PS* associated with grand coalitions). Also intuitively, these deleterious effects of *NP* are greatest where parliamentary support is weakest. Subsequently, we address the statistical uncertainty of these estimates.

Our approaches for interpreting interaction terms also apply when the interacted variables have been nonlinearly transformed, such as squared terms (a special case of linear interaction where a variable in essence interacts with itself so that its effect depends on its own level), higher order polynomials, and logs. Such nonlinear transformations also render interpretation of estimated *effects* from simple examination of estimated *coefficients* very difficult and again highlight the utility of differentiation or differencing for interpreting regression analyses employing interaction terms.

Consider the case when a researcher believes that the effect of some variable, *x*, depends on the level of that variable *x*. One way to model

TABLE 6.   Predicted *Government Duration*

	NP = 1	NP = 2	NP = 3	NP = 4
PS = 40	33.05	20.42	7.79	−4.84
PS = 50	31.87	23.93	15.99	8.05
PS = 60	30.70	27.44	24.18	20.93
PS = 70	29.52	30.95	32.38	33.81
PS = 80	28.34	34.46	40.57	46.69

Note: Predicted values are calculated at given values, setting *PD* = 1.

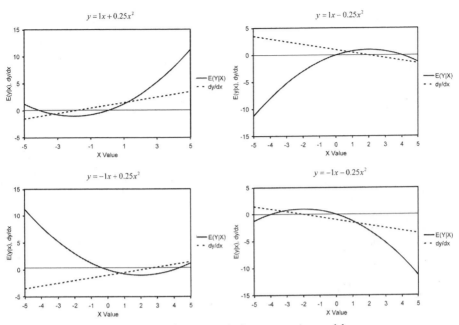

Fig. 1. Quadratic terms in linear-regression models

this proposition is to include a quadratic, or squared term, x^2, that is, the interaction of x with itself. Researchers have applied this type of interaction in several domains: the effect of age is modeled as quadratic in studies of political participation; the effect of time elapsed or remaining is modeled as quadratic in studies of the dynamics of political campaigns; loss functions in many rational-choice models take quadratic form, and so on. Generically, such quadratic models might appear as

$$y = \beta_0 + \beta_{x1}x + \beta_{x2}x^2 + \varepsilon \tag{18}$$

and specify parabolic (hump-shaped, convex or concave) relations of x and y. As always, the effect of x on y can be calculated through differentiation as

$$\partial y/\partial x = \beta_{x1} + 2\beta_{x2}x \tag{19}$$

or by differencing predicted values of y as x moves from x_a to x_c:

$$\hat{y}_c - \hat{y}_a = \hat{\beta}_0 + \hat{\beta}_{x1}x_c + \hat{\beta}_{x2}x_c^2 - (\hat{\beta}_0 + \hat{\beta}_{x1}x_a + \hat{\beta}_{x2}x_a^2)$$
$$= \hat{\beta}_{x1}(x_c - x_a) + \hat{\beta}_{x2}(x_c^2 - x_a^2) \tag{20}$$

Figure 1 demonstrates how these parabolic relationships, and the associated marginal effects, look under the four possible combinations of positive and negative coefficients on the linear and quadratic terms, β_{x1} and β_{x2}, respectively, across the positive and negative value range of x.

The effect of parliamentary support on government duration, for example, might depend on the level of parliamentary support itself in this way. One might well expect an additional 10 percent support to contribute less to extending a government's duration if that increase is from 75 percent to 85 percent than if it is from 45 percent to 55 percent. Table 7 shows the estimation results of a simple model to evaluate this possibility.

Given the signs of the coefficients on PS and PS^2, negative and positive, respectively, and the strictly positive values of PS, ranging from about 40 percent to 80 percent, this example will resemble the lower left quadrant of figure 1. Figure 2 plots the estimated effect line (calculated using (19)) and predicted government duration as a function of PS. (Substantively, the estimated relationship seems odd and intriguing.)

Another commonly used nonlinear transformation is the natural logarithm, $\ln(x)$, which is often used when researchers want to allow the marginal effect of x to decline at higher levels of x as we have suggested here regarding the effect of parliamentary support on government duration. Common examples include the natural logs of dollars (e.g., budgetary outlays, gross domestic product [GDP], campaign spending, or personal income), of population or population density, or of elapsed time (e.g., milliseconds for response latencies or other units such as hours or

TABLE 7. OLS Regression Results, *Government Duration:* Quadratic-Term Model

	Coefficient (standard error) p-Value
Parliamentary Support (PS)	−2.734
	(2.061)
	0.200
Parliamentary Support, squared (PS²)	0.0257
	(0.017)
	0.142
Intercept	95.20
	(62.44)
	0.144
N (df)	22 (19)
Adjusted R^2	0.158
$P > F$	0.075

Note: Cell entries are the estimated coefficient, with standard error in parentheses, and two-sided p-level (probability $|T| > t$) referring to the null hypothesis that $\beta = 0$ in italics.

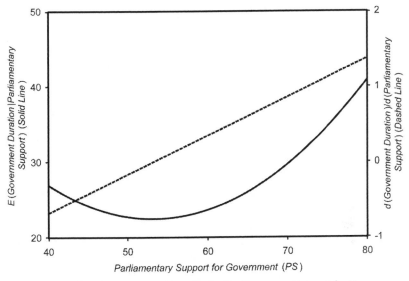

Fig. 2. Predicted *Government Duration* by *Parliamentary Support for Government,* quadratic model

days). In each of these cases, researchers will often expect the marginal effect of a unit increase in x to be greater at lower values of x and to diminish as x itself increases. In a linear-additive model, the log of x relates linearly to y, while x relates nonlinearly to y. The log transformation can also be included in the linear-interactive model. Consider, for example, a model including the natural log of parliamentary support interacted with the number of governing parties, controlling for party discipline:

$$Government\ Duration = \beta_0 + \beta_{np}NP + \beta_{\ln(ps)}\ln(PS) + \beta_{np\ln(ps)}NP$$
$$\times\ (\ln(PS)) + \beta_{pd}PD + \varepsilon \qquad (21)$$

Table 8 gives the estimation results for this model. The slightly higher adjusted R^2 of this model and the generally greater significance of its coefficient estimates compared with the model in table 5 suggest that this model with diminishing government-duration returns to parliamentary support is somewhat superior. The effect of parliamentary support on government duration in this model can be calculated by differentiating with respect to PS: $\partial GD/\partial PS = (\beta_{\ln(ps)} + \beta_{np\ln(ps)}NP)/PS$.[20] Differentiating

---

20. Recall that $\partial(\ln x)/\partial x = 1/x$ and the chain rule for nested functions specifies $\partial f(g(x))/\partial x = \partial f(g)/\partial g \times \partial g(x)/\partial x$. For a model, $y = \beta_0 + \beta_{\ln(x)}\ln(x) + \beta_z z + \beta_{\ln(x)z}\ln(x) \times z$, the marginal effect of x is

$$\partial y/\partial x = (\beta_{\ln(x)} + \beta_{\ln(x)z}z)(\partial\ln(x)/\partial x) = (\beta_{\ln(x)} + \beta_{\ln(x)z}z)(1/x)$$

Please see appendix A for further description of these differentiation rules.

TABLE 8. OLS Regression Results, *Government Duration*: Log-Transformation Interactive Model

	Coefficient (standard error) p-Value
Number of Parties (NP)	−136.97
	(48.984)
	0.012
ln(*Parliamentary Support*) (ln(*PS*))	−43.410
	(27.417)
	0.132
Number of Parties × ln(*Parliamentary Support*) (NP × ln(PS))	32.710
	(11.956)
	0.014
Party Discipline (PD)	9.960
	(3.172)
	0.006
Intercept	201.41
	(111.16)
	0.088
N (*df*)	22 (17)
Adjusted R^2	0.520
$P > F$	0.002

Note: Cell entries are the estimated coefficient, with standard error in parentheses, and two-sided p-level (probability $|T| > t$) referring to the null hypothesis that $\beta = 0$ in italics.

with respect to *NP* yields its conditional effect on government duration: $\partial GD/\partial NP = \beta_{np} + \beta_{npln(ps)} \ln(PS)$.

Note that, befitting the diminishing returns specified for *PS*, the predicted values will vary depending upon the values of *PS* selected for the calculations. Differences in predicted values also remain straightforward to calculate.[21] Figure 3 shows one informative way to present these estimation results, plotting the predicted government duration as a function of parliamentary support at a few substantively revealing levels of the number of governing parties. (Party discipline is held fixed at one (high)

---

21. Specifically, the difference in predicted values of $\hat{y}$, as x increases from x_a to x_c, is

$$\hat{y}_c - \hat{y}_a = \hat{\beta}_0 + \hat{\beta}_{\ln(x)} \ln(x_c) + \hat{\beta}_z z + \hat{\beta}_{\ln(x)z} \ln(x_c) \times z - (\hat{\beta}_0 + \hat{\beta}_{\ln(x)} \ln(x_a) + \hat{\beta}_z z$$
$$+ \hat{\beta}_{\ln(x)z} \ln(x_a) \times z)$$
$$= \hat{\beta}_{\ln(x)} \ln(x_c) - \hat{\beta}_{\ln(x)} \ln(x_a) + \hat{\beta}_{\ln(x)z} \ln(x_c) \times z - \hat{\beta}_{\ln(x)z} \ln(x_a) \times z$$
$$= \hat{\beta}_{\ln(x)} \ln(x_c/x_a) + \hat{\beta}_{\ln(x)z} \ln(x_c/x_a) \times z$$
$$= \ln(x_c/x_a)(\hat{\beta}_{\ln(x)} + \hat{\beta}_{\ln(x)z} z)$$

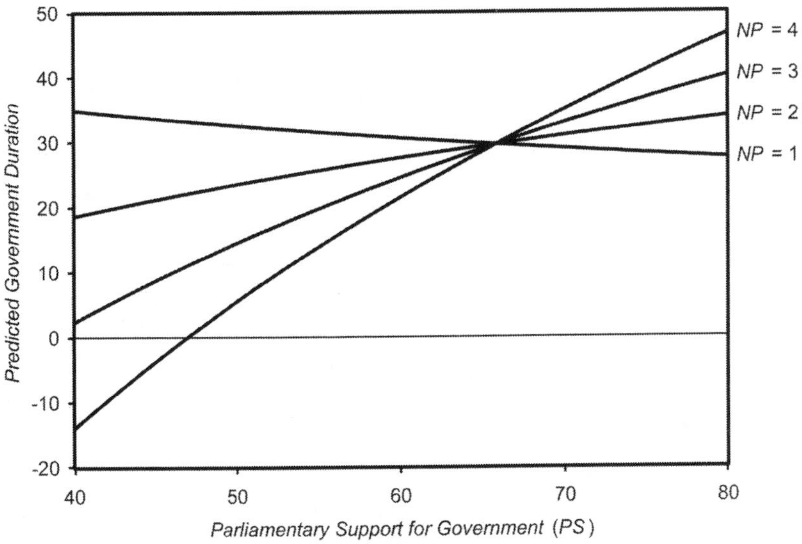

Fig. 3.   Predicted *Government Duration* by *Parliamentary Support for Government,* log-transformation model

in fig. 3.) The figure reveals the essentially flat relationship between parliamentary support and government duration for single-party governments; the generally deleterious effects of the number of parties in government, especially at lower levels of parliamentary support; and the increasing effect of parliamentary support at higher numbers of governing parties. It also reveals the diminishing-returns relation of parliamentary support to government duration imposed by the log transformation. This concavity becomes more noticeable when the number of governing parties is greater, that is, when the effects of support are greater.

Threshold and spline (a.k.a. slope-shift) models represent another class of independent-variable transformations commonly used in combination with interaction terms. Suppose a researcher thought that the effect of some independent variable x changed sign or magnitude beyond some particular value, x_0. For example, the effect of years of education, *YE*, on a person's income, *Inc*, might shift at certain numbers of years representing the passing of key milestones, say, at sixteen years (typical college graduation). Up to that point, the accumulated years represent prebaccalaureate education; beyond it they represent some branch of advanced professional training. One way to specify an empirical model reflecting such a proposition would be to create a new indicator variable, call it *PB* for postbaccalaureate, equal to one if $YE \geq 16$ and zero if $YE < 16$. To allow the effect of *YE* to shift at year sixteen and above, we would want to interact YE with this transformation of itself, PB, to yield the following model:

$$Inc = \beta_0 + \beta_1 YE + \beta_2 YE \times PB + \beta_3 PB + \varepsilon \tag{22}$$

This model has a discontinuity at $YE = 16$, and so using the difference method will prove more intuitive. (In fact, the function is not differentiable at $YE = 16$.) From the difference method, then, we see exactly how the effect of income in this model of adding a year of education depends on whether that year is one of the first fifteen, the sixteenth, or beyond the sixteenth.

For values of $YE < 15$ (where an additional year of schooling would not activate the threshold of PB), a one-unit shift in schooling from YE_a to YE_c would imply a β_1 shift in income:

$$\Delta Inc = \beta_0 + \beta_1 YE_c + \beta_2 YE_c \times 0 + \beta_3 \times 0 - (\beta_0 + \beta_1 YE_a$$
$$+ \beta_2 YE_a \times 0 + \beta_3 \times 0)$$

$$\Delta Inc = \beta_1$$

For values of YE such that $15 \leq YE < 16$ (where the additional year of schooling activates the threshold of PB), a one-unit shift in schooling from YE_a to YE_c would imply a $\beta_1 + \beta_2 YE_c + \beta_3$ shift in income:

$$\Delta Inc = \beta_0 + \beta_1 YE_c + \beta_2 YE_c \times 1 + \beta_3 \times 1 - (\beta_0 + \beta_1 YE_a$$
$$+ \beta_2 YE_a \times 0 + \beta_3 \times 0)$$

$$\Delta Inc = \beta_1 + \beta_2 YE_c + \beta_3$$

For values of $YE \geq 16$ (where the additional year of schooling does not change the value of PB), a one-unit shift in schooling would imply a $\beta_1 + \beta_2$ shift in income:

$$\Delta Inc = \beta_0 + \beta_1 YE_c + \beta_2 YE_c \times 1 + \beta_3 \times 1 - (\beta_0 + \beta_1 YE_a$$
$$+ \beta_2 YE_a \times 1 + \beta_3 \times 1)$$

$$\Delta Inc = \beta_1 + \beta_2$$

In this slope-shift or threshold model, the prebaccalaureate piece of the income-education relation may not adjoin the postbaccalaureate piece; rather, a discontinuous jump may occur at the point. To force the segments to link continuously requires a spline model that simply regresses income on YE and $YE^* = YE - 16$ for $YE \geq 16$ and 0 otherwise. This general approach to slope-shift model specification and interpretation extends intuitively to any number of discontinuous or splined-continuous slope shifts (see Greene 2003, secs. 7.2.4–75, pp. 120–22, for further discussion).[22]

Differentiation and/or differencing thus render calculation of the estimated effects of x on y straightforward in any linear-regression model, however the independent variables may have been transformed and in whatever combinations they may interact. The section "Nonlinear Models" in chapter 5 discusses interpretation of interaction terms in nonlinear models, in which these same techniques apply.

Chained, Three-Way, and Multiple Interactions

Interactions involving more than two variables are also possible, of course, and may often be suggested theoretically. Generically, the effect of some x on y could depend on two (or more) other variables, w and z (etc.), as in this model:

$$y = \beta_0 + \beta_x x + \beta_z z + \beta_w w + \beta_{xz} xz + \beta_{xw} xw + \varepsilon \qquad (23)$$

By differentiation, the effects of x, w, and z are $\partial y/\partial x = \beta_x + \beta_{xz} z + \beta_{xw} w$, $\partial y/\partial w = \beta_w + \beta_{xw} x$, and $\partial y/\partial z = \beta_z + \beta_{xz} x$, respectively. In our government-duration analysis, for example, one might well conjecture that party discipline, that is, parties' internal strategic unity, would as likely moderate the effects of the number of governing parties on government duration as would parliamentary support. A linear-interactive specification that could entertain this possibility would be

$$\textit{Government Duration} = \beta_0 + \beta_{np} NP + \beta_{ps} PS + \beta_{pd} PD$$
$$+ \beta_{npps} NP \times PS + \beta_{nppd} NP \times PD + \varepsilon \quad (24)$$

Interpretation of estimated conditional effects can once again proceed equally by differences or derivatives: $\partial GD/\partial NP = \beta_{np} + \beta_{npps} PS + \beta_{nppd} PD$, $\partial GD/\partial PS + \beta_{ps} + \beta_{npps} NP$, and $\partial GD/\partial PD = \beta_{pd} + \beta_{nppd} NP$, again safely ignoring the binary nature of PD in deriving these expressions for the conditional effects (but remembering it when considering at what values of PD or for what magnitude change in PD to calculate those conditional effects). This sort of asymmetric model, in which one variable (here NP) modifies the effects of several others (here PD and PS) or, equivalently, has its effect modified by several others (perhaps the

---

22. The model could equivalently be expressed as $Inc = \beta_0 + \beta_1 YE \times (1 - PB) + \beta_2 YE \times PB + \beta_3 PB + \varepsilon$. The one-unit shift at $YE < 15$ would still imply a β_1 shift in income. The one-unit shift at $15 \leq YE < 16$ would imply a $-\beta_1 YE_a + \beta_2 YE_c + \beta_3$ shift in income, and the one-unit shift at $YE \geq 16$ would imply a β_2 shift in income.

more intuitive way to express it in this substantive case), but in which those others do not condition each other's effects, might be termed a "chained-interaction" model.[23]

Substantively in this example, a model in which NP has its effects on government duration moderated by PD and PS certainly makes sense, but the first column of results in table 9 gives little empirical support for this chained specification, compared with the simpler model in table 5. However, we might also expect PD and PS, the missing pairwise interaction, to condition each other's government-duration effects. The durability benefits of extra seats of parliamentary support should logically depend on the reliability of those seats' votes for the government, that is, on party discipline. We call an empirical model like the one this suggests, in which the effect of each variable depends on each of the others, the complete "pairwise-interaction" model, which here just adds that one further interaction term, $PD \times PS$, to the model:

$$Government\ Duration = \beta_0 + \beta_{np}NP + \beta_{ps}PS + \beta_{pd}PD$$

$$+ \beta_{npps}NP \times PS + \beta_{nppd}NP \times PD$$

$$+ \beta_{pdps}PD \times PS + \varepsilon \qquad (25)$$

Differentiation, as always, suffices to calculate the conditional effects in this model:

$$\frac{\partial GD}{\partial NP} = \beta_{np} + \beta_{npps}PS + \beta_{nppd}PD, \qquad \frac{\partial GD}{\partial PS} = \beta_{ps} + \beta_{npps}NP + \beta_{pdps}PD,$$

$$\frac{\partial GD}{\partial PD} = \beta_{pd} + \beta_{nppd}NP + \beta_{pdps}PS$$

Table 9 also presents the estimation results for this pairwise-interaction model, which has stronger empirical support, although the difficulty of estimating this many coefficients,[24] especially on such correlated regressors, in just twenty-two observations is also becoming evident in the standard errors of those coefficient estimates (perhaps not so much or in the same way regarding the estimated effects but we are deferring for now the discussion of the statistical certainty of conditional-effect estimates).

Finally, one might push even further along these lines to suggest that not only should the effect of each of these three factors depend on each

---

23. We thank an anonymous reviewer for suggesting this name for such models.

24. A model with k unique independent variables and all their pairwise interactions will comprise $(k!)/2(k-2)! + k$ regressors.

of the others in all pairwise interactions, but the effect of each might logically depend on the combination of the others present as well. For example, the government-durability benefit of additional seats of parliamentary support certainly should depend on the reliability of those seats' votes, and so we are theoretically and substantively rather confident of the $PS \times PD$ interaction. However, the impact of this "reliability adjusted" additional parliamentary support on government duration might then depend on the number of governing parties by the same logic that led us to our initial model with its single interaction term, $PS \times NP$. Table 9 also gives the estimation results for such a "fully interactive" model, which adds $PS \times NP \times PD$ to the set of pairwise interactive terms. Obviously, we are now straining the available information in the mere twenty-two observations of our example data set severely, but the empirical support

TABLE 9.   OLS Regression Results, *Government Duration:* Three-Way Interactive Models

	Chained-Interaction Model	Pairwise-Interaction Model	Fully Interactive Model
Number of Parties (NP)	−33.810	−27.766	−51.265
	(12.013)	(11.535)	(41.342)
	0.012	*0.029*	*0.235*
Parliamentary Support (PS)	−0.66773	−1.5115	−2.0949
	(0.47518)	(0.61940)	(1.1699)
	0.179	*0.028*	*0.095*
Party Discipline (PD)	14.859	−48.690	−86.847
	(7.758)	(33.670)	(72.969)
	0.073	*0.169*	*0.254*
Number of Parties × Parliamentary Support (NP × PS)	0.52785	0.43443	0.84262
	(0.20651)	(0.1970)	(0.7171)
	0.021	*0.043*	*0.260*
Number of Parties × Party Discipline (NP × PD)	−2.6514	−3.4973	22.233
	(3.7263)	(3.4716)	(43.531)
	0.487	*0.330*	*0.617*
Party Discipline × Parliamentary Support (PD × PS)		1.1624	1.8219
		(0.60174)	(1.2709)
		0.073	*0.174*
Number of Parties × Parliamentary Support × Party Discipline (NP × PS × PD)			−0.44313
			(0.74719)
			0.563
Intercept	62.191	108.039	141.495
	(27.159)	(34.545)	(66.556)
	0.036	*0.007*	*0.052*
N (df)	22 (16)	22 (15)	22 (14)
Adjusted R^2	0.4967	0.5701	0.5507
P > F	0.0053	0.0031	0.0069

Note: Cell entries are the estimated coefficient, with standard error in parentheses, and two-sided *p*-level (probability $|T| > t$) referring to the null hypothesis that $\beta = 0$ in italics.

for this fully interactive model over the preceding pairwise-interactive model seems weak at any rate.[25]

Interpretation of estimated effects in such highly interactive models from coefficient estimates alone would be especially problematic. For example, the coefficient β_{np} in each of these models refers to the effect of NP when both PS and PD are zero, and the former, of course, is logically impossible. Using the derivative method allows for better interpretation:

$$\frac{\partial GD}{\partial NP} = \begin{cases} -33.81 + 0.528(PS) - 2.651(PD) \\ \quad \text{in the chained-interaction model} \\ -27.77 + 0.434(PS) - 3.497(PD) \\ \quad \text{in the pairwise-interaction model} \\ -51.26 + 0.843(PS) + 22.23(PD) - 0.443(PS \times PD) \\ \quad \text{in the fully interactive model} \end{cases}$$

$$\frac{\partial GD}{\partial PS} = \begin{cases} -0.6677 + 0.528(NP) \\ \quad \text{in the chained-interaction model} \\ -1.511 + 0.434(NP) + 1.162(PD) \\ \quad \text{in the pairwise-interaction model} \\ -2.095 + 0.843(NP) + 1.822(PD) - 0.443(NP \times PD) \\ \quad \text{in the fully interactive model} \end{cases}$$

$$\frac{\partial GD}{\partial PD} = \begin{cases} 14.86 - 2.651(NP) \\ \quad \text{in the chained-interaction model} \\ -48.69 - 3.497(NP) + 1.162(PS) \\ \quad \text{in the pairwise-interaction model} \\ -86.85 + 22.23(NP) + 1.822(PS) - 0.443(NP \times PS) \\ \quad \text{in the fully interactive model} \end{cases}$$

The conditional effects of each independent variable in a three-way (multiple) interaction model, excepting the variables not chained in a chained-interaction model, depend on the values of two (or more) other independent variables. Accordingly, effective interpretation will require the presentation of three (or more) dimensions of information: the value of each of the conditioning variables and the estimated conditional effect corresponding to those values. The section "Presentation of Interactive Effects" in this chapter provides useful strategies for doing this.

In summary, these exercises in interpretation of coefficients should underscore the point that the variables in interactive specifications have

---

25. A model with k unique independent variables and all possible unique interactions of all subsets (including the whole set) of those k factors will comprise $2^k - 1$ regressors.

varying effects. The size and sign of the effect of x can depend critically upon the value at which the other variable, z, is held; conversely, the size and sign of the effect of z can depend critically upon the value at which the other variable, x, is held. Calling one of the coefficients involved in these effects the "main effect" and another the "interactive effect" can be quite misleading and is no substitute for understanding the model's actual estimated effects. Outside the purely linear-additive model, coefficients are not effects. Differentiation and differences of predicted values are two simple, universally applicable, and reliable tools for examining the effect of variables x and z on y in general and in interactive models in particular.

Once we have calculated these estimated conditional effects, however, we must also estimate and convey the statistical certainty of those estimates. We next discuss how to calculate standard errors for estimated conditional effects (as opposed to coefficients) and determine the degree to which these effects (as opposed to coefficients) are statistically distinguishable from zero.

Linking Statistical Tests with Interactive Hypotheses

Common social-science practice in testing interactive propositions relies almost exclusively on t-tests of significance of individual coefficients in the model. Researchers commonly compare each of the three key coefficient estimates in a typical interactive model, for example, $\hat{\beta}_x$, $\hat{\beta}_z$, and $\hat{\beta}_{xz}$ in the standard linear-interactive model, (14), to its respective standard error. Assuming that the model exhibits the necessary statistical properties otherwise (i.e., the Gauss-Markov conditions), the ratios in this comparison are t-distributed (or asymptotically normal), and so these tests are statistically valid (asymptotically). However, scholars often mistake their meaning—that is, they often mistake what these t-tests actually test—reflecting the persistent confusion of coefficients for effects and the use of the misleading main- and interactive-effect terminology. Just as the effects of variables involved in interactive terms depend upon two (or more) *coefficients* and the values of one (or more) other variable(s), so too do judgments of uncertainty surrounding those effects: their standard errors and the relevant t-statistics, confidence intervals, and hypothesis-test results (significance levels).

Single t-tests on individual coefficients on variables involved in interactive terms require care to interpret because they refer to significance at only one empirical value of the other variables. For example, β_x and β_z in our standard model (8) indicate, respectively, x's effect on y *when z*

equals zero and z's effect on y *when x equals zero,* and so the standard
t-tests on our estimates $\hat{\beta}_x$ and $\hat{\beta}_z$ indicate the significance of that variable's effect when the other variable equals zero. These specific values of
zero, as noted before, may be substantively, empirically, or even logically
irrelevant.

For example, in our model of the number of presidential candidates,
the number of ethnic groups never equals zero in the sample and, logically, could not. Likewise in the government-duration example, neither
the number of governing parties nor the level of parliamentary support
could ever be zero. Thus, any inferences drawn about the statistical sig-
nificance of β_R, the coefficient on *Runoff* in table 1, or of β_{np}, β_{ps}, or β_{pd}
in any of the models of table 9, are largely meaningless because they refer
to conditions that could not logically exist. On the other hand, inferences
drawn about the statistical significance of our estimate of coefficient β_G
in table 2 refer to the impact of *Groups* in the substantively meaningful
case where *Runoff* equals zero. With no runoff system, the number of
ethnic groups decreases the number of presidential candidates, and the
test of whether the decrease is statistically significantly distinguishable
from zero (i.e., no change) produces a p-value of 0.228.

Likewise in our model of U.S. support for social welfare (table 3), the
coefficients on *Female* and *Republican* each refer to substantively important conditions. The term $\hat{\beta}_F$ is the gender gap when *Republican*
equals zero, that is, among Democrats, which is substantively tiny
(-0.003) and statistically indistinguishable from zero (i.e., insignificant,
at $p = 0.828$), whereas $\hat{\beta}_R$ is the partisan gap when *Female* equals zero
(among males), which is substantively sizable (0.22) and highly statistically distinguishable from zero $(p < 0.001)$.

Even in cases like these last three, however, where individual coefficients refer to logically possible conditions that exist in the sample and,
indeed, have important substantive meaning, the judgment of statistical
significance is still a limited one. In the first case, it applies only to the effect of *Groups* in the absence of runoffs $(Runoff = 0)$ and says nothing
about that effect where runoffs occur $(Runoff = 1)$. In the latter two
cases, the t-tests on the interaction terms refer only to the significance of
the gender gap among Democrats and to the partisan gap among males,
and they say nothing of the other two gaps (the gender gap among Republicans and the partisan gap among females). Moreover, the specific
conditions to which the coefficient estimates and their estimated standard errors refer have no greater claim than the remaining conditions do
to being "main" effects in any sense.

To provide a universally valid framework for hypothesis testing of effects rather than coefficients in interactive models, consider the following types of theoretical questions often asked about them: (1) Does y depend on x, or, equivalently, is y a function of x? Does y depend on z, or, equivalently, is y a function of z? (2) Is y's dependence on x *contingent* upon or *moderated* by z, or, equivalently, does the effect of x on y depend on z? Is y's dependence on z *contingent* upon or *moderated* by x, or, equivalently, does the effect of z on y depend on x? **This is the classic interactive hypothesis; the two sets of questions are logically identical.** (3) Does y depend on x, z, and/or their interaction, xz, at all, or, equivalently, is y a function of x, z, and/or xz? In tables 10–12, we link each of these sets of theoretical questions about interactive relationships to hypotheses, and each hypothesis to its mathematical expression and to its correspondingly appropriate statistical test(s).

We start with the simpler propositions in table 10. Note that the statistical test that corresponds to each hypothesis states a null hypothesis that, as always, is what the researcher would like, theoretically, to reject statistically. The first hypothesis examines whether x has any effect on y. The mathematical expression for testing the effect of x on y includes β_x and $\beta_{xz}z$. The standard F-test on the pair of coefficients, β_x and β_{xz}, therefore identifies whether x matters (i.e., whether y depends on x). Only these coefficients both being zero would imply that y does not depend on x in any fashion in this model.

An extension of this first hypothesis would propose some direction to the effect of x on y. The "simple" extension that the effect of x on y is positive or negative is actually ill defined in linear-interactive models because the effects of x vary linearly depending on values of z, implying

TABLE 10. Does y Depend on x or z?

Hypothesis	Mathematical Expression	Statistical Test
x affects y, or y is a function of (depends on) x	$y = f(x)$ $\partial y/\partial x = \beta_x + \beta_{xz}z \neq 0$	F-test $H_0: \beta_x = \beta_{xz} = 0$
x increases y	$\partial y/\partial x = \beta_x + \beta_{xz}z > 0$	Multiple t-tests: $H_0: \beta_x + \beta_{xz}z \leq 0$
x decreases y	$\partial y/\partial x = \beta_x + \beta_{xz}z < 0$	Multiple t-tests: $\beta_x + \beta_{xz}z \geq 0$
z affects y, or y is a function of (depends on) z	$y = g(z)$ $\partial y/\partial z = \beta_z + \beta_{xz}x \neq 0$	F-test: $H_0: \beta_z = \beta_{xz} = 0$
z increases y	$\partial y/\partial z = \beta_z + \beta_{xz}x > 0$	Multiple t-tests: $H_0: \beta_z + \beta_{xz}x \leq 0$
z decreases y	$\partial y/\partial z = \beta_z + \beta_{xz}x < 0$	Multiple t-tests: $H_0: \beta_z + \beta_{xz}x \geq 0$

Note: Table assumes standard linear-interactive model, $y = \gamma_0 + \beta_x x + \beta_z z + \beta_{xz}xz + \varepsilon$, is specified.

that the effects will be positive for some, equal to (and around) zero for some, and negative for other z values, although, as stressed before, not all values of z will necessarily be substantively relevant. Accordingly, no common practice exists for testing hypotheses that x or z generally increases or decreases y in linear-interactive models because hypotheses like these are logically ambiguous in such models. Depending on where the relevant ranges of z lie, and on the accompanying standard errors, the effects could therefore be significantly positive in some meaningful ranges, significantly negative in others, and indistinguishable from zero in yet others.

To illustrate, suppose we hypothesize that x has an increasingly positive effect on y as z increases, starting from no effect at $z = 0$. Suppose also that $z < 0$ is logically impossible. In this case, even if that proposition were true and the evidence strongly supported it, the estimated effect of x on y would be zero at $z = 0$ and therefore necessarily statistically indistinguishable from zero at that point. The estimated effect also must be statistically indistinguishable from zero for some range near $z = 0$, given that all estimates have some error. (Obviously, the insignificant range will be larger the less precisely the relevant coefficients are estimated.) Therefore, hypotheses that the effects of x (or z) are generally positive or negative should instead be specified over some range of z (or x).

In stating hypotheses that prescribe the range of values of the conditioning variable(s) over which they are to be evaluated, researchers should calculate measures of uncertainty to determine whether the effects of x at several specific values of z are statistically distinguishable from zero. This approach is highlighted in the second and third hypotheses in table 10. Then, to evaluate a claim that the effect of x on y is generally positive or negative, the researcher could test whether the effect of x on y is positive over the entire logically possible, or substantively sensible, or sample range of z by conducting several t-tests along the range of z.[26] Alternatively, but equivalently, he or she could plot $\partial \hat{y}/\partial x$ over an appropriate range of z along with confidence intervals. These confidence intervals would indicate rejection of the null hypothe-

---

26. Alternatively, the researcher could simply estimate a linear-additive model that omits the interaction in question and test whether the coefficient on x or z significantly differs from zero in the usual manner. If the interaction truly exists, the linear-additive model would tend to produce for coefficients on x and z their *average* effect across the sample values of the other variable. If the interaction does truly exist, however, the researcher must note that this linear-additive model is misspecified, with the coefficient estimates on x and z therefore likely subject to attenuation bias and inefficiency. Accordingly, these tests would tend to be biased toward failing to reject.

sis at all values of z where zero lies outside the confidence interval around the estimated effect. However, as just explained, researchers must recognize that, in some cases, we would expect failure to reject (confidence intervals that span zero) at some levels of z even if the hypothesis generally were very strongly supported by the data.[27]

To execute this set of t-tests or generate these confidence intervals, the researcher will first need to calculate the estimated conditional effect by the differentiation or difference method. In (14), for instance, the marginal effect of x on y is $\partial\hat{y}/\partial x = \hat{\beta}_x + \hat{\beta}_{xz}z$. As always, to express the uncertainty of an estimated effect, in standard errors or in confidence intervals around it, we must find its variance. It is critical to note that the coefficient estimates vary across repeated samples, not the values of z; that is, the estimated coefficients are the random variables, whereas z is fixed.[28] The estimated effect of x on y contains the product of $\hat{\beta}_{xz}$ and z; correspondingly, the estimated conditional effects will have some level of uncertainty that depends on z. Just as the effects of x on y vary with the values of z, the standard errors of the effects of x on y also vary with values of z. Each unique value in the set of estimated conditional effects (one at each value of z) will have its own variance and corresponding standard error.[29] The variance of $\partial\hat{y}/\partial x$, the estimated marginal effect of x on y, is[30]

---

27. Recognizing this issue, we suggest subsequently that researchers plot the estimated effects of x across meaningful ranges of z, along with confidence intervals, and then consider the share of these confidence intervals' covered area that lies above (or below) zero as an indication of how strongly the evidence supports the proposition. Since "generally" is imprecise and involves judgment, this test is imprecise and involves judgment too, but visualizing graphically the proportion of a confidence area that lies above or below zero should help in rendering this judgment.

28. Recall that the classical linear-regression model assumes that z is fixed in repeated sampling or that, if z is stochastic, we interpret our estimates as conditioning on z (i.e., given z or holding z constant). Either way, in our estimated effects, z is fixed; $\hat{\beta}$ is what varies due to estimation uncertainty.

29. One must distinguish between the variance of the estimated marginal effect of x on y given z, $V(\partial E(y|x,z)/\partial x)$; the variance of the estimated effect of a discrete change in x on y given z, $V(\Delta E(y|x,z)/\Delta x)$; the variance of the prediction or estimate itself, $V[E(y|x,z)]$; and the variance of the prediction or forecast error, $V[y - E(y|x,z)]$. Both estimation error in $\hat{\beta}$ and the stochastic residual or error term in the model, ε, arise in the fourth case (variance of the prediction or forecast error). The variances of estimates and of estimated effects, that is, all of the other cases, involve only the estimation error in $\hat{\beta}$.

30. Given some constant c and some random variable r, $V(cr) = c^2V(r)$. Given some constant c and two random variables r_1 and r_2, the variance of the expression $V(r_1 + cr_2)$ $= V(r_1) + c^2V(r_2) + 2cC(r_1,r_2)$. In our context, the x and z are fixed in repeated sampling, per the standard OLS assumptions, and the estimated coefficients are the random variables. More generally, for a vector of random variables, $\hat{\beta}$, and a constant vector, $\mathbf{m}$, the variance of the linear-additive function $\mathbf{m}'\hat{\beta}$ is $V(\mathbf{m}'\hat{\beta}) = \mathbf{m}'V(\hat{\beta})\mathbf{m}$. Expression (26) is just one specific example of this more general formula.

$$V(\partial\hat{y}/\partial x) = V(\hat{\beta}_x) + z^2 V(\hat{\beta}_{xz}) + 2zC(\hat{\beta}_x, \hat{\beta}_{xz}) \tag{26}$$

Our uncertainty regarding the conditional effects of x on y thus depends on variability in our estimates of β_x and β_{xz}, the covariance between those estimates of β_x and β_{xz}, and the values of z at which the effects are evaluated. Our estimates of $V(\hat{\beta}_x)$ and $V(\hat{\beta}_{xz})$ are simply the squares of the standard errors of the coefficient estimates, $\hat{\beta}_x$ and $\hat{\beta}_{xz}$, reported in typical regression output. The covariance of $\hat{\beta}_x$ and $\hat{\beta}_{xz}$, however, is not typically displayed in standard regression output. It must be extracted from the estimated variance-covariance matrix of the coefficient estimates.

A variance-covariance matrix[31] is a symmetric matrix that contains the variance of each estimated coefficient along the diagonal elements and the covariance of each estimated coefficient with the other estimated coefficients in the off-diagonal elements:

$$\mathbf{V}(\hat{\boldsymbol{\beta}}) = \begin{bmatrix} V(\hat{\beta}_1) & & & \\ C(\hat{\beta}_1,\hat{\beta}_2) & V(\hat{\beta}_2) & & \\ \vdots & & \ddots & \\ C(\hat{\beta}_1,\hat{\beta}_k) & C(\hat{\beta}_2,\hat{\beta}_k) & \cdots & V(\hat{\beta}_k) \end{bmatrix}$$

In practice, we use estimates of $V(\hat{\beta}_x)$, $V(\hat{\beta}_{xz})$, and $C(\hat{\beta}_x, \hat{\beta}_{xz})$, which we will designate as $\widehat{V(\hat{\beta}_x)}$, $\widehat{V(\hat{\beta}_{xz})}$, and $\widehat{C(\hat{\beta}_x,\hat{\beta}_{xz})}$. The desired estimate of $C(\hat{\beta}_x,\hat{\beta}_{xz})$ will appear as the off-diagonal element in the estimated variance-covariance matrix that corresponds to $\hat{\beta}_x$ and $\hat{\beta}_{xz}$. In most software, researchers can easily retrieve this estimated variance-covariance matrix by a single postestimation command.[32]

To execute the tests or construct the confidence intervals suggested in the second and third rows of table 10, then, the researcher calculates the effect of x at some value of z, $\partial\hat{y}/\partial x = \hat{\beta}_x + \hat{\beta}_{xz}z$, and the estimated variance around that effect at that value of z, $\widehat{V(\partial\hat{y}/\partial x)}$. The t-statistic for testing whether this estimate is statistically distinguishable from zero is then found by dividing the estimated effect $\partial\hat{y}/\partial x$ by the estimated standard error of $\partial\hat{y}/\partial x$ and evaluating the result against the t-distribution (with $n - k$ degrees of freedom, with n the number of observations and k the number of regressors, including the constant). The researcher would then repeat the process for other values across the relevant range of z to determine whether a general claim can be made about the direction of the effect.

---

31. In OLS, the variance-covariance matrix of the estimated coefficient vector is $s^2(\mathbf{X'X})^{-1}$, where s^2 is our estimate of σ^2, the variance of ε.

32. In STATA, this command is "vce".

As suggested earlier, however, determining whether the effect of x on y is *generally, typically,* or *on-average* positive or negative, a common component of the typical complex of interactive hypotheses, requires more precise definition of the italicized terms. If *on-average* refers to the effect at the sample-average value of z, then the single t-test of the effect of x at that value of z suffices. This value of z also gives the appropriate estimated effect and its statistical confidence for an on-average effect taken to mean the average in the sample of the effect of x.[33] If, however, one wishes to gauge the statistical certainty surrounding the hypothesis that the effect of x on y is generally or typically positive, we suggest plotting the $\partial \hat{y}/\partial x$ over the sample range of z, with confidence intervals.[34] Support for the hypotheses that $\partial y/\partial x$ is generally or typically positive or negative would correspond to most (unfortunately, no firm cutoff share exists) of this confidence interval lying appropriately above or below zero. One could quantify the share of the area covered by the confidence interval that lies above or below zero to give more precision to this analysis.[35]

Aside from these basic hypotheses that x affects y (perhaps with some sign over some range of z), researchers are also interested in whether and how the effects of x and of z on y depend on the other variable. Table 11 presents these interactive hypotheses.

Notice that the coefficient on xz directly reflects the presence, sign, and substantive magnitude of this conditioning relationship: that is, the degree to which the effects of x and z on y depend on the other variable's value. As such, the standard t-test of the coefficient on the multiplicative term tests for the presence or sign of a conditioning relationship. Since the effect of x on y is $\partial y/\partial x = \beta_x + \beta_{xz}z$, a simple t-test of the null hypothesis that $\beta_{xz} = 0$ directly evaluates whether the effect of x changes as z changes. A rejection of the null hypothesis that $\beta_{xz} = 0$ thus supports the most central interactive hypothesis: the effect of x on y varies with the level of z (and vice versa). If interactive hypotheses contain a directional

---

33. The first section of chapter 4 shows that this hypothesis also corresponds to the standard t-statistic reported for the coefficient on x^* in an interactive model where x and z have been mean-centered (had their sample means subtracted) to x^* and z^*.

34. The "Presentation of Interactive Effects" section in this chapter discusses how to construct confidence intervals.

35. An alternative strategy would be to estimate a different model, one without the interaction term(s), and simply evaluate the usual t-test on the appropriate coefficient, on x or on z. This alternative would reveal directly whether, on average or generally, x or z had a nonzero effect on y. However, if the true relationship really is interactive, then this alternative model is misspecified, and so these t-tests would be, at minimum, inefficient. See note 27.

TABLE 11.   Is y's Dependence on x Conditional on z and Vice Versa? How?

Hypothesis	Mathematical Expression	Statistical Test
The effect of x on y depends on z	$y = f(xz, \cdot)$	t-test: H_0: $\beta_{xz} = 0$
	$\partial y/\partial x = \beta_x + \beta_{xz}z = g(z)$	
	$\partial(\partial y/\partial x)/\partial z = \partial^2 y/\partial x \partial z = \beta_{xz} = 0$	
The effect of x on y increases in z	$\partial(\partial y/\partial x)/\partial z = \partial^2 y/\partial x \partial z = \beta_{xz} > 0$	t-test: H_0: $\beta_{xz} \leq 0$
The effect of x on y decreases in z	$\partial(\partial y/\partial x)/\partial z = \partial^2 y/\partial x \partial z = \beta_{xz} < 0$	t-test: H_0: $\beta_{xz} \geq 0$
The effect of z on y depends on x	$y = f(xz, \cdot)$	t-test: H_0: $\beta_{xz} = 0$
	$\partial y/\partial z = \beta_z + \beta_{xz}x = h(x)$	
	$\partial(\partial y/\partial z)/\partial x = \partial^2 y/\partial z \partial x = \beta_{xz} = 0$	
The effect of z on y increases in x	$\partial(\partial y/\partial z)/\partial x = \partial^2 y/\partial z \partial x = \beta_{xz} > 0$	t-test: H_0: $\beta_{xz} \leq 0$
The effect of z on y decreases in x	$\partial(\partial y/\partial z)/\partial x = \partial^2 y/\partial z \partial x = \beta_{xz} < 0$	t-test: H_0: $\beta_{xz} \geq 0$

Note: Table assumes standard linear-interactive model, $y = \gamma_0 + \beta_x x + \beta_z z + \beta_{xz}xz + \varepsilon$, is specified.

element—for example, the effect of x on y increases as z increases, or the effect of x on y decreases as z increases—researchers might apply one-tailed tests of the null hypothesis that $\beta_{xz} \leq 0$ or $\beta_{xz} \geq 0$. These directional hypotheses are displayed in the second and third lines of table 11.[36]

Note, also, that the mathematical expression and the statistical test for the hypothesis that x conditions the effect of z on y are identical to those for the converse that z conditions the effect of x on y. This reflects the logical symmetry of interactive propositions. If z conditions the effect of x on y, then x logically must condition the effect of z on y and in the same amount. In fact, the second three rows of table 11 simply state the logical converses of the first three rows, and so the corresponding mathematical expressions and statistical tests are identical.[37]

Finally, table 12 reveals the statistical test corresponding to the broadest sort of hypothesis one might have regarding an interactive model: that y depends in some manner, be it in a linear-additive and/or a linear-interactive way, on x and/or on z. In common language, some one or combination of x and z matters for y. This corresponds statistically, quite simply, to the F-test that all three coefficients involved in the inter-

---

36. Since assuming directionality in this way lowers the empirical hurdle for statistical rejection, many scholars opt more conservatively for always employing nondirectional hypotheses and two-tailed tests.

37. The order of differentiation in a cross-derivative never matters, and so this symmetry does not rely on the linear-multiplicative form specifically. In any logical proposition/mathematical model, that the effect of x depends on z implies that the effect of z depends, in identical fashion, on x: $\partial(\partial y/\partial x)/\partial z \equiv \partial(\partial y/\partial z)/\partial x$ for any function $y(x,z)$. In this case, the effect of x on y, or how y changes as x changes, is $\partial y/\partial x = \beta_x + \beta_{xz}z$. The effect of z on that effect of x on y, or how z changes the effect of x on y, is analogously $\partial(\partial y/\partial x)/\partial z = \partial(\beta_x + \beta_{xz}z)/\partial z = \beta_{xz}$. The converses for the effect of z on y and how x modifies this effect are $\partial y/\partial z = \beta_z + \beta_{xz}x$ and $\partial(\partial y/\partial z)/\partial x = \partial(\beta_z + \beta_{xz}x)/\partial x = \beta_{xz}$.

TABLE 12.   Does *y* Depend on *x*, *z*, or Some Combination Thereof?

Hypothesis	Mathematical Expression	Statistical Test
y is a function of (depends on) *x*, *z*, and/or their interaction	$y = f(x,z,xz)$	*F*-test: H_0: $\beta_x = \beta_z = \beta_{xz} = 0$

Note: Table assumes standard linear-interactive model, $y = \gamma_0 + \beta_x x + \beta_z z + \beta_{xz} xz + \varepsilon$, is specified.

action, β_x, β_z, β_{xz}, are zero. That all three of these are zero is the only condition that would render *x* and *z* wholly irrelevant to *y*.

Let us walk our first empirical example through the tests outlined in tables 10–12.

First, does *x* affect *y*? Does the number of presidential candidates depend in some linear or linear-interactive way on the number of ethnic groups? An *F*-test of the null hypothesis that $\beta_G = 0$ and $\beta_{GR} = 0$ addresses this question. The *F*-test produces these results:[38] $F = 2.62$; $\text{Prob}(F_{2,12} > 2.62) = 0.1140$. Whether to reject the null hypothesis depends on the researcher's desired level of certainty. At conventional levels ($p < 0.10$, $p < 0.05$, $p < 0.01$), the researcher would not (quite) reject the null.[39]

Does *z* affect *y*? Does the number of presidential candidates depend in some linear or linear-interactive way on the presence of a runoff system? The *F*-test of the null hypothesis that $\beta_R = 0$ and $\beta_{GR} = 0$ yields the following results: $F = 2.96$; $\text{Prob}(F_{2,12} > 2.96) = 0.0903$, which would (barely) satisfy a $p < 0.10$ criterion but fail the stricter $p < 0.05$, $p < 0.01$ criteria.

Next, we ask whether *x* (generally) increases *y*. To answer this question, the researcher should conduct *t*-tests of or construct confidence intervals for the effect of *x* across some range of values of *z* (corresponding to "generally"). To conduct these *t*-tests, one must first calculate the standard errors associated with the given marginal effect following equation (26). Table 13 displays the estimated variance-covariance matrix from our example, which we will need for these calculations.[40]

The element in the first row and first column, 0.593, is the estimated

---

38. In our notation, *F* is the calculated *F*-statistic, and $\text{Prob}(F_{n,m} > F)$ is the probability, under the null, of a value greater than *F* in an *F*-distribution with *n* and *m* degrees of freedom; that is, the *p*-level at which the null is rejected.

39. A less strictly classical approach to hypothesis testing would simply report the *p*-level and leave the reader to determine how much weight to assign a result with this level of statistical significance.

40. Appendix B provides step-by-step STATA commands for conducting these types of calculations.

variance of $\hat{\beta}_G$, which is the square of its standard error from table 1: $0.770^2 \approx 0.593$. Likewise, the estimated variance of $\hat{\beta}_{GR}$ is the square of its standard error reported in table 1: $0.941^2 \approx 0.885$. The information we need from the variance-covariance matrix that we do not see in the typical regression output is $\widehat{C(\hat{\beta}_G, \hat{\beta}_{GR})}$, which is -0.593. To calculate the estimated variance of the estimated marginal effects, we simply substitute these values from the estimated variance-covariance matrix into equation (26).

$$\widehat{V(\partial\hat{y}/\partial G)} = \widehat{V(\hat{\beta}_G)} + Runoff^2 \; \widehat{V(\hat{\beta}_{GR})} + 2 \times Runoff$$
$$\times \; \widehat{C(\hat{\beta}_G, \hat{\beta}_{GR})}$$

$$\widehat{V[(\partial\hat{y}/\partial G) \mid Runoff = 0]} = 0.593 + 0^2 \times 0.885 + 2 \times 0 \times -0.593$$
$$= 0.593$$

$$\widehat{V[(\partial\hat{y}/\partial G) \mid Runoff = 1]} = 0.593 + 1^2 \times 0.885 + 2 \times 1 \times -0.593$$
$$= 0.292$$

The proposition that societal groups increase the number of presidential candidates corresponds to the null hypothesis: H_0: $\beta_G + \beta_{GR} Runoff \leq 0$. This null hypothesis can be evaluated at the two valid values of z: zero (no runoff system) and one (a runoff system). Table 14 gives these results.

With a one-tailed p-value of 0.884, we cannot reject the null hypothesis that $\beta_G + \beta_{GR} Runoff \leq 0$ when $Runoff = 0$. In systems without runoffs, a negative or null relationship between *Groups* and *Candidates* cannot be rejected. However, with a one-tailed p-value of 0.041, we *can*

TABLE 13. Estimated Variance-Covariance Matrix of Coefficient Estimates, Predicting *Number of Presidential Candidates*

	Groups	Runoff	Groups × Runoff	Intercept
Groups	0.593			
Runoff	0.900	2.435		
Groups × Runoff	−0.593	−1.377	0.885	
Intercept	−0.900	−1.509	0.900	1.509

TABLE 14. Hypothesis Tests of whether *Groups* Affects *Number of Presidential Candidates*

	$\partial\hat{y}/\partial G$	s.e. $(\partial\hat{y}/\partial G)$	t-Statistic	One-Tailed p-Value H_0: $\beta_G + \beta_{GR}Runoff \leq 0$	One-Tailed p-Value H_0: $\beta_G + \beta_{GR}Runoff \geq 0$	90% Confidence Interval
Runoff = 0	−0.979	0.770	−1.271	0.886	0.114	[−2.352, 0.394]
Runoff = 1	1.026	0.540	1.902	0.041	0.959	[0.064, 1.988]

reject the null hypothesis that *Groups* decrease or have no effect on *Candidates* in favor of the alternative that some positive relationship between *Groups* and *Candidates* exists when *Runoff* = 1. To test the reverse directional hypothesis, that the number of societal groups decreases the number of presidential candidates, we pose the opposite null hypothesis: $\beta_G + \beta_{GR}Runoff \geq 0$ and reevaluate. In the absence of a runoff system, the one-tailed *p*-value is 0.116, which actually (substantively oddly, as we have noted) approaches significance. In the presence of a runoff system, the one-tailed *p*-value of 0.959 suggests that we are quite unable to reject the null hypothesis of a positive or null relationship between *Groups* and *Candidates*.

To test the analogous directional hypotheses with respect to the effect of a runoff system on the number of presidential candidates, the researcher could conduct a number of *t*-tests over a logically relevant range of *Groups*. Table 15 displays some examples.

Here, we see that evaluation of the null hypothesis of $\beta_R + \beta_{GR}Groups \leq 0$ changes for various values of *Groups*. As the number of ethnic groups increases, our ability to reject the null hypothesis that runoff systems reduce the number of candidates increases. When *Groups* exceeds 1.5, the hypothesis test begins to approach conventional significance levels. At *Groups* = 2, we can reject the null hypothesis that runoff systems reduce the number of candidates. To investigate whether a runoff system decreases the number of presidential candidates, we reevaluate the *t*-statistics for the null hypothesis: $\beta_R + \beta_{GR}Groups \geq 0$. At the resulting one-tailed *p*-values, we cannot remotely reject the null hypothesis in any case, thus lending no support at all to the reverse proposition.

So far, then, the evidence perhaps weakly suggests that the number of ethnic groups relates to the number of presidential candidates and slightly less weakly suggests that the presence or absence of runoff systems does so. The best that might be said regarding the results for the general direction of these relationships is that the evidence suggesting that runoffs

TABLE 15. Hypothesis Tests of whether *Runoff* Affects *Number of Presidential Candidates*

	$\partial\hat{y}/\partial R$	s.e. $(\partial\hat{y}/\partial R)$	*t*-Statistic	One-Tailed *p*-Value H_0: $\beta_R + \beta_{GR}Groups \leq 0$	One-Tailed *p*-Value H_0: $\beta_R + \beta_{GR}Groups \geq 0$	90% Confidence Interval
Groups = 1	−0.486	0.752	−0.646	0.735	0.265	[−1.826, 0.854]
Groups = 1.5	0.517	0.542	0.954	0.180	0.820	[−0.449, 1.483]
Groups = 2	1.520	0.682	2.229	0.023	0.977	[0.305, 2.735]
Groups = 2.5	2.522	1.038	2.430	0.016	0.984	[0.672, 4.373]
Groups = 3	3.525	1.461	2.413	0.016	0.984	[0.922, 6.128]

and ethnic fragmentation might generally decrease the number of presidential candidates is consistently and considerably weaker than the evidence weighing in the other, the theoretically expected, direction.

Next, continuing to the tests outlined in table 11, comes the question of whether the effect of *Groups* on *Candidates* depends in some way on the presence or absence of a runoff system and vice versa. The answer to this central substantive question of interactive models emerges directly from the coefficient on the interactive term, β_{GR}, and its standard error. A two-tailed test of the null hypothesis H_0: $\beta_{GR} = 0$ yields a *p*-value of 0.054. Determination of "statistical significance" depends on the researcher's acceptable level of uncertainty: rejection at the $p < 0.10$ threshold, near rejection at a $p < 0.05$ threshold, and failure to reject at the tighter $p < 0.01$ level. The symmetry of interaction terms also implies the same answer for whether *Groups* modifies the effect of *Runoff*.

The directional hypothesis of whether runoffs increase the effect of *Groups* on *Candidates* requires a one-tailed test of the null H_0: $\beta_{GR} \leq 0$, which yields a *p*-value of 0.027. The researcher can reject the null hypothesis of a negative or nonzero coefficient in favor of the alternative hypothesis of some positive coefficient at the 0.10 and 0.05 levels but not at the 0.01 level. The positive effect of *Groups* on *Candidates* does seem larger in runoff systems, and *Runoff* has greater positive effect with a higher number of *Groups*.

Finally, consider the test in table 12: whether *x* and *z* have any effect on *y* in some linear or linear-interactive fashion. Here, the researcher cares whether *Groups*, *Runoff*, and/or their product affects *Candidates*. An *F*-test that all three coefficients are zero, H_0: $\beta_G = \beta_R = \beta_{GR} = 0$, yields the following results: $F = 2.27$, with a *p*-value from the $F_{3,12}$ distribution of 0.132: not overwhelming, but not surprising and perhaps not too disappointing either, given the small sample size.

We consider the remaining empirical examples more quickly. In the support for social welfare example, an *F*-test on the coefficients on *Female* and the interaction between *Female* and *Republican* addresses the interesting substantive question of whether gender affects support for social welfare. The results, $F = 13.08$; $\text{Prob}(F_{2,1073} > 13.08) = 0.000$, allow us to reject confidently the null hypothesis that gender has no effect on support for social welfare. Analogously, the *F*-test of the two coefficients on *Republican* and on the interaction of *Female* and *Republican* produces $F = 144.07$; $\text{Prob}(F_{2,1073} > 144.07) = 0.000$, allowing confident rejection of the null hypothesis that partisanship has no effect on support for social welfare. Next, we test whether the effect of gender depends on partisanship and vice versa. Recall that these calculations require access to values in the estimated variance-covariance matrix. Table 16 contains these values.

Table 17 shows that the statistical significance of the effect of gender varies sharply by partisanship: among Democrats, we can reject neither of the directional hypotheses (no statistically discernible effect of gender exists among Democrats). Among Republicans, in contrast, we can reject the null hypothesis that females are less supportive of social welfare, at $p < 0.001$. Table 18 considers the converse: whether partisanship affects support for social welfare at various values of *Female*, that is, among males and among females. Here, the null hypothesis that *Republican* increases support for social welfare is soundly rejected among both females and males: being a Republican significantly decreases support for social welfare. Finally, an *F*-test of all three coefficients addresses

TABLE 16. Estimated Variance-Covariance Matrix of Coefficient Estimates, Predicting *Support for Social Welfare*

	Female	Republican	Female × Republican	Intercept
Female	0.00021			
Republican	0.00012	0.00024		
Female × Republican	−0.00021	−0.00024	0.00046	
Intercept	−0.00012	−0.00012	0.00012	0.00012

TABLE 17. Hypothesis Tests of whether *Female* Affects *Support for Social Welfare*

	$\partial \hat{y}/\partial F$	s.e. $(\partial \hat{y}/\partial F)$	t-Statistic	One-Tailed p-Value H_0: $\beta_F +$ β_{FR}Republican ≤ 0	One-Tailed p-Value H_0: $\beta_F +$ β_{FR}Republican ≥ 0	95% Confidence Interval
Republican = 0	−0.003	0.0144	−0.218	0.586	0.414	[−0.031, 0.025]
Republican = 1	0.081	0.0158	5.109	0.000	0.999	[0.050, 0.111]

TABLE 18. Hypothesis Tests of whether *Republican* Affects *Support for Social Welfare*

	$\partial \hat{y}/\partial R$	s.e. $(\partial \hat{y}/\partial R)$	t-Statistic	One-Tailed p-Value H_0: $\beta_R +$ β_{FR}Female ≤ 0	One-Tailed p-Value H_0: $\beta_R +$ β_{FR}Female ≥ 0	95% Confidence Interval
Female = 0	−0.220	0.0155	−14.18	0.999	0.000	[−0.251, −0.190]
Female = 1	−0.137	0.0147	−9.33	0.999	0.000	[−0.166, −0.108]

whether partisanship or gender affect support for social welfare some-how. Here $F = 103.65$; $\text{Prob}(F_{3,1073} > 103.65) = 0.000$, and so we can confidently conclude that gender, partisanship, and/or their interaction significantly predict support for social welfare.

In our simple government-duration example (table 5), we test the hypothesis that parliamentary support for government has an effect on government duration using an F-test of the coefficients on PS and the interaction of PS and NP: $F = 6.50$; $\text{Prob}(F_{2,17} > 6.50) = 0.008$; we can confidently reject the null hypothesis of no effect. Similarly, the F-test of the coefficients on NP and $NP \times PS$ identifies whether the number of governing parties has an effect on government duration: $F = 4.87$; $\text{Prob}(F_{2,17} > 4.87) = 0.021$; we can reject the null hypothesis of no effect at conventional significance levels of $p < 0.10$ and $p < 0.05$.

The proposition that the number of governing parties decreases governmental duration must be evaluated at particular values of PS. The estimated variance-covariance matrix is provided in table 19. Table 20 gives the relevant calculations. When parliamentary support ranges from 40 percent to 60 percent, we can reject the null hypothesis that the number of governing parties increases governmental duration at conventional levels. However, when parliamentary support is high (at 70 percent), we

TABLE 19.   Estimated Variance-Covariance Matrix of Coefficient Estimates, Predicting *Government Duration*

	Number of Parties	Parliamentary Support	NP × PS	Party Discipline	Intercept
Number of Parties (NP)	128.712				
Parliamentary Support (PS)	4.564	0.206			
NP × PS	−2.089	−0.078	0.035		
Party Discipline	2.980	0.080	−0.058	10.265	
Intercept	−274.906	−11.870	4.587	−10.666	699.881

TABLE 20.   Hypothesis Tests of whether *Number of Parties* Affects *Government Duration*

	$\partial \hat{y}/\partial NP$	s.e. $(\partial \hat{y}/\partial NP)$	t-Statistic	One-Tailed p-Value $H_0: \beta_{np} + \beta_{npps}PS \leq 0$	One-Tailed p-Value $H_0: \beta_{np} + \beta_{npps}PS \geq 0$	90% Confidence Interval
PS = 40	−12.628	4.135	−3.054	0.996	0.004	[−19.822, −5.434]
PS = 50	−7.942	2.558	−3.104	0.997	0.003	[−12.393, −3.492]
PS = 60	−3.257	1.711	−1.903	0.963	0.037	[−6.233, −0.280]
PS = 70	1.429	2.500	0.572	0.288	0.712	[−2.920, 5.778]
PS = 80	6.115	4.063	1.505	0.075	0.925	[−0.954, 13.183]

can reject neither of the directional hypotheses, and when support is extremely high, we can actually weakly reject (at $p < 0.075$) the null hypothesis that the number of governing parties has the theoretically expected negative effect on government duration. Conversely for the effect of parliamentary support, table 21 shows that with only one governing party, neither directional hypothesis is rejected; greater parliamentary support might increase, decrease, or have no effect upon the duration of single-party governments. However, with multiple governing parties, the null hypothesis that parliamentary support decreases government duration is rejected. Thus, generally, parliamentary support seems to enhance government durability as expected, although we cannot reject the alternative for the case of single-party governments. Finally, the hypothesis that the number of governing parties, governing support, and/or their interaction significantly affects duration of governments can be evaluated using an F-test of all three coefficients. This F-test produces $F = 4.62$; $\text{Prob}(F_{3,17} > 4.62) = 0.015$. We can confidently reject the null hypothesis of no effect.

In table 7, we considered a simple model in which parliamentary support had a nonlinear relation to government duration, specified empirically by including PS and PS^2 as regressors. In this case, the effect of PS on government duration is $\partial GD/\partial PS = \beta_{ps} + 2\beta_{ps^2}PS$. The test that PS has some effect on government duration is the F-test of both coefficients, for which the table reports $p = 0.075$: moderate support. The test of whether this effect depends (linearly) on the level of parliamentary support itself (i.e., that the relationship of PS to government duration would be quadratic) is the standard t-test on $\hat{\beta}_{ps^2}$, which is reported in the table as giving $p = 0.142$: weak support. To gauge the significance of the estimated effect of PS at particular values of PS, we would use the following formula:

TABLE 21. Hypothesis Tests of whether *Parliamentary Support* Affects *Government Duration*

	$\partial \hat{y}/\partial PS$	s.e. $(\partial \hat{y}/\partial PS)$	t-Statistic	One-Tailed p-Value $H_0: \beta_{ps} + \beta_{npps}NP \leq 0$	One-Tailed p-Value $H_0: \beta_{ps} + \beta_{npps}NP \geq 0$	90% Confidence Interval
$NP = 1$	−0.118	0.293	−0.402	0.654	0.346	[−0.627, 0.392]
$NP = 2$	0.351	0.185	1.897	0.037	0.963	[0.029, 0.673]
$NP = 3$	0.820	0.228	3.587	0.001	0.999	[0.422, 1.217]
$NP = 4$	1.288	0.374	3.448	0.002	0.998	[0.638, 1.938]

$$\widehat{V\left(\frac{\partial GD}{PS}\right)} = \widehat{V(\hat{\beta}_{ps} + 2\hat{\beta}_{ps^2}PS)}$$

$$= \widehat{V(\hat{\beta}_{ps})} + \widehat{V(2\hat{\beta}_{ps^2}PS)} + 2\widehat{C(\hat{\beta}_{ps},2\hat{\beta}_{ps^2}PS)}$$

$$= \widehat{V(\hat{\beta}_{ps})} + 4PS^2 \times \widehat{V(\hat{\beta}_{ps^2})} + 2PS \times 2\widehat{C(\hat{\beta}_{ps},\hat{\beta}_{ps^2})}$$

$$= \widehat{V(\hat{\beta}_{ps})} + 4PS^2 \times \widehat{V(\hat{\beta}_{ps^2})} + 4PS \times \widehat{C(\hat{\beta}_{ps},\hat{\beta}_{ps^2})} \qquad (27)$$

The relevant portion of the estimated variance-covariance matrix of these coefficient estimates is

$$\widehat{V(\hat{\beta}_{ps})} \approx 4.247 \qquad \widehat{C(\hat{\beta}_{ps^2},\hat{\beta}_{ps})} \approx -0.0343$$

$$\widehat{C(\hat{\beta}_{ps},\hat{\beta}_{ps^2})} \approx -0.0343 \qquad \widehat{V(\hat{\beta}_{ps^2})} \approx 0.000281$$

So, for example, the standard error of the estimated marginal effect of PS on government duration at $PS = 55$ percent is

$$s.e.(\widehat{\partial GD/\partial PS}) = \sqrt{4.247 + 4 \times 55^2 \times 0.000281 + 4 \times 55 \times -0.0343}$$

$$\approx 0.3$$

The marginal effect at this point is $-2.73 + 2(0.0257)55 = +0.09$ and, given the associated standard error of the marginal effect, is not remotely statistically distinguishable from zero. In fact, the estimated marginal effect is insignificant in one- or two-tailed tests over about half of the sample range of PS in this model; to present the range over which the marginal effect is distinguishable from zero, we suggest calculating and plotting the confidence intervals around the effect line depicted in figure 2 (as we do in fig. 10). We provide instructions for doing so in the next section.

In table 8, we log-transformed parliamentary support before including it in an interactive model otherwise identical to that of table 5. Accordingly, testing null hypotheses that effects equal zero (i.e., testing for the existence of effects) follow that discussion exactly. The variable NP has no effect on government duration if and only if (*iff*) the coefficients on NP and $NP \times \ln(PS)$ are both zero ($F = 5.36$; $\text{Prob}(F_{2,17} > 5.36) = 0.0157$: reject); PS has no effect *iff* the coefficients on $\ln(PS)$ and $NP \times \ln(PS)$ are both zero ($F = 6.78$; $\text{Prob}(F_{2,17} > 6.78) = 0.0068$: reject); and NP and PS have no linear or linear-interactive effect *iff* all three coefficients are zero ($F = 4.81$; $\text{Prob}(F_{3,17} > 4.81) = 0.0133$: reject). The significance of the estimated marginal effects of NP at specific values of $\ln(PS)$ and the test of whether NP generally decreases government duration likewise follow the discussion from the table 5 case exactly, merely replacing PS with $\ln(PS)$. However, estimated marginal effects of PS on government duration are $\partial \widehat{GD}/\partial PS = (\hat{\beta}_{\ln(ps)} + \hat{\beta}_{np\ln(ps)}NP)/PS$, which depend on both NP and PS; so too, then, does the standard error of this estimated effect:

$$
\begin{aligned}
\widehat{V(\partial \widehat{GD}/\partial PS)} &= \widehat{V((\hat{\beta}_{\ln(ps)} + \hat{\beta}_{np\ln(ps)}NP)/PS)} \\
&= \overline{V((\hat{\beta}_{\ln(ps)} + \hat{\beta}_{np\ln(ps)}NP)/PS)} \\
&= \frac{1}{PS^2}(\widehat{V(\hat{\beta}_{\ln(ps)})} + \widehat{V(\hat{\beta}_{np\ln(ps)}NP)} + 2\widehat{C(\hat{\beta}_{\ln(ps)},\hat{\beta}_{np\ln(ps)}NP)}) \\
&= \frac{1}{PS^2}(\widehat{V(\hat{\beta}_{\ln(ps)})} + NP^2\,\widehat{V(\hat{\beta}_{np\ln(ps)})} + 2NP \\
&\quad \times \overline{C(\hat{\beta}_{\ln(ps)},\hat{\beta}_{np\ln(ps)})})
\end{aligned} \tag{28}
$$

We simply insert the values from the estimated variance-covariance matrix of these coefficient estimates, along with assigned values of NP and PS, into this formula for the variance of the marginal effect of PS at those values of PS and NP.[41] For example, a three-party government that increased its parliamentary support marginally from 55 percent would increase its expected duration by about a month ($\partial \widehat{GD}/\partial PS = -43.4/55 + (32.7 \times 3)/55 \approx 0.995$), with a standard error for that estimate of $[(1/55)^2\,751.7 + (3/55)^2\,142.95 - (2 \times 3)/55^2 \times 302.8]^{0.5} \approx 0.271$. Dividing the estimated marginal effect by the estimated standard error yields a t-statistic of $0.995/0.271 = 3.671$, implying reject at $p(t_{17} > 3.671) = 0.0019$, for the test of the null hypothesis of no effect of PS on government duration at these levels of NP and PS. As with the preceding nonlinear transformation, however, we strongly recommend graphical presentation of such estimated effects and confidence intervals and so defer further discussion.

For the chained, pairwise, and fully interactive three-way-interaction models of government duration (table 9), finally, we could follow the same sequence of common theoretical hypotheses. In doing so, notice first that we can evaluate whether one of the three independent variables affects the dependent variable by conducting an F-test of the null hypothesis that the coefficients on all of the terms involving that variable are zero. For example, the F-test that PS "matters" has a null hypothesis that β_{ps} and β_{npps} are both zero in the chained model ($F = 5.86$; $p(F_{2,16} > 5.86) = 0.012$); that β_{ps}, β_{npps}, and β_{pspd} are zero in the pairwise model

---

41. Here $\widehat{V(\hat{\beta}_{\ln(ps)})}$ and $\widehat{V(\hat{\beta}_{np\ln(ps)})}$ are the squares of the standard errors reported in table 8. The term $\overline{C(\hat{\beta}_{\ln(ps)},\hat{\beta}_{np\ln(ps)})}$ is obtained by calling up the variance-covariance matrix (not shown), $\overline{C(\hat{\beta}_{\ln(ps)},\hat{\beta}_{np\ln(ps)})} = -302.8$.

$(F = 5.81; p(F_{3,15} > 5.81) = 0.008)$; and that β_{ps}, β_{npps}, β_{pspd}, and β_{nppspd} are all zero in the fully interactive model $(F = 4.26; p(F_{4,14} > 4.26) = 0.018)$. That the effect of PS depends on NP or PD in the pairwise or fully interactive models is now also a joint hypothesis in the pairwise and fully interactive models: that the coefficients β_{npps} and β_{pspd} are both zero $(F = 5.69; p(F_{2,15} > 5.69) = 0.015)$ and that β_{npps}, β_{pspd}, and β_{nppspd} are all zero, respectively $(F = 3.75; p(F_{3,14} > 3.75) = 0.036)$. That the effect of PS depends on NP or that the effect of PS depends on PD are both simple-hypothesis t-tests in the pairwise model, on β_{npps} or β_{nppd} ($t = 2.2$, $p(|t_{15}| > 2.2) = 0.04$; $t = 1.9$, $p(|t_{15}| > 1.9) = 0.07$), respectively, but each is a joint-hypothesis F-test of β_{npps} and β_{nppspd} or of β_{pspd} and β_{nppspd} $(F = 2.5; p(F_{2,14} > 2.5) = 0.12; F = 1.96; p(F_{2,14} > 1.96) = 0.18)$, respectively, in the fully interactive model. The tests for the analogous hypotheses regarding how the effects of NP or of PD depend on the one other variable or the two other variables are symmetric. Finally, that some linear or linear-interactive combination of NP, PS, and/or PD "matters" corresponds to the F-test of the model in each case (as reported in table 9: $F = 4.68$; $p(F_{7,14} > 4.68) = 0.007$). We highly recommend graphical methods for interpreting the sign and the statistical certainty and significance of estimated effects of each variable over ranges of each of the others, as discussed in the next section.

Presentation of Interactive Effects

Hayduk and Wonnacott (1980) noted, "While the technicalities of these [interactive] procedures have received some attention . . . the proper methods for the interpretation and visual presentation of regressions containing interactions are not widely understood" (400). This section provides guidance on presentation of results from models that include interaction terms.

Mere presentation of regression coefficients and their standard errors is inadequate for the interpretation of interactive effects. As we have seen, the estimated effects of variables involved in interactive terms and the standard errors of these estimated effects vary depending on the values of the conditioning variables. Therefore, conditional effects, as best calculated by the derivative or difference method, are most effectively conveyed in tabular and graphical forms. In the political-science literature, presentations of effects that involve interactive terms now often do utilize tables or graphs that depict the effect of x on y when z equals particular values. Presentation of estimated conditional effects across a sufficiently wide and meaningful range of values of z and indication of the estimated uncertainty of these estimated conditional effects across that range are still too often lacking, however.

Many statistical-software packages can provide conditional marginal effects or predicted values as well as standard errors for these conditional estimates, typically as part of some postestimation suite of commands.[42] Further, other programs exist that will generate estimates of uncertainty around predicted values from any estimated model using simulation (King, Tomz, and Wittenberg 2000). While we have no particular qualms about such preprogrammed commands and procedures,[43] the procedures we recommend maximize the user's control over the values at which marginal effects and predicted values are calculated and, we believe, will strengthen the user's understanding and intuition in interpreting models that contain interactive terms. We strongly recommend that the user be fully conversant with the elementary mathematical foundations underlying these procedures before taking preprogrammed commands "off the shelf."[44]

Presentation of Marginal Effects

Researchers will often wish to convey to the reader how the effect of x changes over some range of z values. The estimated marginal conditional effects of x on y are the first derivative of $\hat{y}$ with respect to x: $\partial \hat{y}/\partial x = \hat{\beta}_x + \hat{\beta}_{xz} z$. We will want to discuss these conditional effects of x over some substantively revealing range of z values. One such revealing range and sequence of values, which may serve as a good default, would be an evenly spaced range of values ranging from a, the sample minimum of z, to c, its sample maximum. More generally, the researcher could calculate

---

42. For example, in STATA, the postestimation command lincom will report estimates, standard errors, t-statistics, p-levels, and a 95 percent confidence interval for any linear combination of coefficients. Appendix B contains syntax that will apply lincom across a range of values.

43. We emphasize, however, that the researcher should verify that the uncertainty estimates produced by these procedures do not, as some unfortunately do, erroneously add stochastic error to estimation error in calculating the uncertainty of estimated effects in models with additively separable stochastic components (like linear regression).

44. This strong warning is especially important when interpreting the effects of interactive variables. Preprogrammed commands that produce marginal effects of variables of interest will likely not recognize that a set of the variables is interactive. As such, these commands may generate a marginal effect for some covariate, naively assuming that all other variables (including the interactive term!) are held constant. This ignores the central fact that the interpretation of the effect of x requires taking into account the coefficient on x, the coefficient on xz, and values of z—underscoring our point that coefficients are not effects in models including interaction terms.

the marginal effect of x on y for any set of z values of interest. Sample means, percentiles, means plus or minus one or two standard errors, and so on, are all frequently useful default points or ranges for these considerations, but substance and researchers' presentational goals should be determinate here. Using z values of particular observations—say, of some well-known, important, or illustrative case or cases in the sample—is also often a good idea. Finally, crucially, the researcher must also report the estimated certainty of the estimated conditional effects in some manner: standard errors, t-statistics, significance levels, confidence intervals. Confidence intervals are usually more effective in graphical presentation and standard errors, t-statistics, or significance levels in tables. Confidence intervals can be generated by this formula:

$$\partial\hat{y}/\partial x \pm t_{df,p} \sqrt{V(\partial\hat{y}/\partial x)}$$

where $t_{df,p}$ is the critical value in a t-distribution with df degrees of freedom ($df = n - k$; n is the number of observations and k the number of regressors, including the constant) for a two-sided hypothesis test at one minus the desired confidence-interval size. For example, to obtain the lower and upper bounds of a 95 percent (90 percent) confidence interval, $t_{df,p}$ should correspond to critical values for a two-sided test at the $p = 0.05$ ($p = 0.10$) level, that is, 0.025 (0.05) on each side; with large degrees of freedom, $t_{df,0.05}$ is approximately 1.96 ($t_{df,0.10} \approx 1.65$).

In our first empirical example, we calculated two sets of conditional effects. We calculated the marginal effect of *Groups* when *Runoff* equals zero and when it equals one, and we calculated the marginal effect of *Runoff* at evenly spaced values of *Groups* from one to three. To construct confidence intervals for these estimated conditional effects, we need to determine the estimated variance of these estimated effects and choose a desired confidence level. Given our small sample size, we choose to accept lower certainty and so select a 90 percent confidence interval. This interval implies a critical value of $t_{12,0.10} = 1.782$. We would thus calculate the upper bound and lower bound for the confidence intervals as

Upper bound: $\partial\hat{y}/\partial x + 1.782 \times \sqrt{V(\partial\hat{y}/\partial x)}$

Lower bound: $\partial\hat{y}/\partial x - 1.782 \times \sqrt{V(\partial\hat{y}/\partial x)}$

Note that $\widehat{V(\partial\hat{y}/\partial x)}$ is the estimated variance of the marginal effect of x on y, produced by plugging in values from the estimated variance-covariance matrix into expression (26). When evaluating the marginal effect of x at several values of z, a graphical display of marginal effects

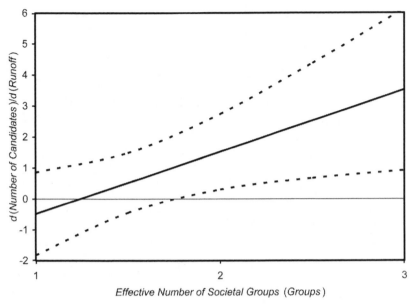

Fig. 4.   Marginal effect of *Runoff*, with 90 percent confidence interval

with confidence intervals is especially effective. Figure 4, for example, displays the marginal effect of *Runoff* across a range of values of *Groups* with confidence intervals around these estimated effects. The straight line thus plots the estimated marginal conditional effects of *Runoff* as a function of *Groups*, and the confidence interval reveals the uncertainty surrounding these estimated effects. The estimated coefficient on the interaction term, $\hat{\beta}_{GR}$, gives our estimate of the slope of the marginal effect line (+2.01), indicating that the marginal effect of *Runoff* on the number of candidates is estimated to increase at a rate of about +2 *Candidates* for each one-unit increase in *Groups*. This graph shows that over the range of sample-relevant values (varying *Groups* from 1 to 3), the marginal effect of *Runoff* increases by about two for each one-unit increase in the number of groups. The marginal effect takes both negative values (though indistinguishable from zero) and positive values along the range of *Groups*. The 90 percent confidence interval overlaps zero at lower values of *Groups*, suggesting that within that range the marginal effect cannot be distinguished from zero statistically, but the confidence interval does not overlap zero when the number of societal groups exceeds 1.75.

Note how this example illustrates the ambiguity discussed previously in hypotheses of "generally positive" effects of variables involved in linear interactions. The researcher in this case would likely have hypothesized that runoff systems increase the number of presidential candidates, especially in more ethnically fragmented societies. Although the effect of runoff systems is essentially zero when fragmentation is very low, this estimated effect turns positive in even moderately fragmented societies, that is, beyond *Groups* ≈ 1.25, and significantly so beyond a modest *Groups* ≈ 1.75. Regarding the proposition that the effect of *Runoff* increases as *Groups* increases, no ambiguity arises. The marginal effect line slopes upward at the rate of $\hat{\beta}_{GR}$, and this estimated slope of the effect line is comfortably significant statistically. The ambiguity arises regarding the hypothesis of a "generally positive" effect, because the estimated effect of *Runoff* is not, in fact, positive over the entire sample range of *Groups* and is only significantly distinguishable from zero in the positive direction over some portions of that sample range. Consideration of only the coefficient on *Runoff*, $\hat{\beta}_R$, would have badly served the researcher in this example; that so-called main-effect coefficient, which actually corresponds to the logically impossible *Groups* = 0 case, is negative and larger than its standard error, yet the actual conditional effects of *Runoff* are indeed estimated to be positive over almost the entire relevant range. Graphing the estimated effects over this substantively relevant range with accompanying confidence intervals in this way reveals that this evidence actually supports that proposition reasonably strongly.

To illustrate the mathematical properties of these effect lines and their associated standard errors, imagine extending the estimated effect line from figure 4 in both directions by projecting into much lower and much higher values for *Groups*. Projecting into values of *Groups* less than 1 is substantively nonsensical, but linear regression per se imposes no such bounds on the values of independent variables, and so let us imagine that it were possible here, solely for these illustrative purposes. Calculating the estimated marginal effects of *Runoff* as the number of ethnic groups ranges from −2 to +6 produces figure 5, demonstrating several interesting properties.

As we noted before, the coefficient on *Runoff* indicates the impact of *Runoff* when *Groups* = 0, and so $\hat{\beta}_R = -2.49$ is also our estimate of the intercept of the marginal effect line (i.e., the value on the y-axis when *Groups* = 0), as the graph indicates. And, as evidenced in figure 4, the estimated coefficient on the interaction term, $\hat{\beta}_{GR}$, gives our estimate of the slope of the marginal effect line (+2.01), indicating that the marginal effect of *Runoff* on the number of candidates is estimated to increase at a rate of about +2 *Candidates* for each one-unit increase in *Groups*. Next, note the hourglass shape of the confidence interval around the estimated marginal-effect line; this hourglass shape is characteristic of confidence intervals for estimated conditional effects in linear-interaction models. The

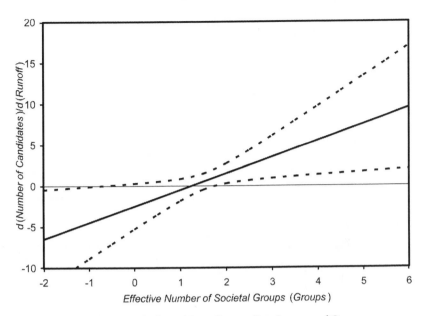

Fig. 5. Marginal effect of *Runoff,* extending the range of Groups

narrowest part of the hourglass occurs at the value of z at which there is greatest certainty concerning the size of the marginal effect of x on y. This point, intuitively, will correspond to the sample mean of the other term in the interaction (z); as always, our estimates have greatest certainty for values around the mean (*centroid* for more than one variable) of our data. The wider parts are points at which lesser certainty prevails regarding the estimated effects, which, intuitively, correspond to points farther from the mean (centroid). The characteristic hourglass shape of the confidence region results from the appearance of z^2 in the expression for the variance of the effect and also from the covariance of the coefficient estimates in that expression, which is typically negative because the corresponding variables x and xz tend to correlate positively. The relative concavity of these hourglasses generally sharpens with the magnitude of this negative correlation. In summary, the confidence intervals (regions) around conditional-effect lines will be (3D) hourglass shaped, with the narrowest points located at the mean (centroid) of the conditioning variable(s) and generally becoming more accentuated as x and xz correlate more strongly, although accentuation depends also on the relative (estimated) variances of $\hat{\beta}_R$ and $\hat{\beta}_{GR}$ and, in appearance, also on graph and z scaling.

Note also from figure 5 that the marginal effect of *Runoff* is statistically distinguishable from zero in the negative direction for values of

Groups below about −0.5, and statistically distinguishable from zero in the positive direction for values of *Groups* above about 1.75. These results illustrate clearly the following points made previously. First, the marginal effect of *Runoff* indeed varies with values of *Groups*. Second, the effect lines, being linear, will extend above and below zero for some (not necessarily meaningful) values of the variables involved. Third, our confidence regarding (i.e., standard errors and significance levels for) the marginal effect of *Runoff* also varies with values of *Groups*. Although figure 5 plots these effects and confidence intervals extending into substantively and even logically meaningless ranges, we emphasize that, in actual research, the researcher bears responsibility to ensure that interpretation and presentation of the results correspond with logically relevant and substantively meaningful values of the independent variables of interest. This implies that researchers must give such information about sample, substantive, and logical ranges necessary for the reader to recognize substantively and logically meaningful and sample-covering ranges. We have projected *Groups* into negative and very high positive values for pedagogical purposes only, to display properties of the marginal effects and confidence intervals most clearly, but we reiterate that these would *not* be logically relevant values in this case. Indeed, presenting a graph like figure 5, which extends well beyond the sample and indeed the logically permissible range, would foster misleading conclusions regarding the substantive meaning of our estimates.

Other types of graphs may more usefully depict marginal effects when conditioning variables are not continuous. For example, the variable *Runoff* takes only two values: zero in the absence of a runoff system and one in the presence of a runoff system. Accordingly, the marginal effect of *Groups* on *Candidates* is also substantively interesting for only these two values of *Runoff*. We can graph the estimated marginal effect of *Groups* on *Candidates* as a function of *Runoff*, as shown in figure 6, with 90 percent confidence intervals around each estimated marginal effect.[45] We see that in systems without runoffs, the confidence interval includes the value of zero, suggesting that the marginal effect of societal groups is not distinguishable from zero in countries with these

---

45. Researchers might also consider plotting normal distributions with means given by the estimated effects and standard deviations by the standard errors of those estimated effects. (Least-squares estimates are at least asymptotically normally distributed thusly.) Another option is a "box-and-whiskers" plot, with the center dots given by the estimated effects, the box around that by a confidence interval or some other multiple of the standard-error range (e.g., plus or minus one standard error), and the whiskers extending to a greater confidence interval or greater multiple of the standard-error range (plus or minus two standard errors). We prefer the simplicity of figures 6 and 7.

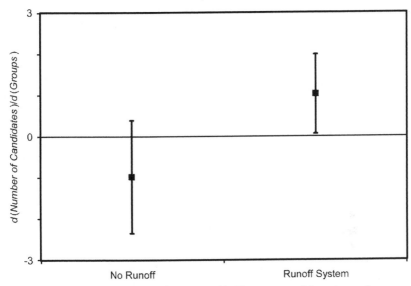

Fig. 6.   Marginal effect of *Groups,* with 90 percent confidence intervals

systems. We also see that in systems with runoffs, the confidence interval does not include the value of zero; the marginal effect of societal groups can be statistically distinguished from zero in these cases. However, the confidence intervals overlap across the values of *Runoff,* suggesting that we cannot say with high levels of certainty that the marginal effects of *Groups* in cases without runoffs and with runoffs are statistically distinguishable from each other.

As another example of using this type of graph, consider our social-welfare example, where both *Female* and *Republican* are dummy variables (binary indicators) and each conditions the other's effect on support for social welfare. Thus, only four effects exist to plot: gender among Democrats and among Republicans and party among women and men. Graphically, conditional effects and associated confidence intervals in such cases are perhaps best displayed as shown in figure 7. (We adopt a more stringent confidence level, 95 percent, in these figures, given the much larger sample here.) Figure 7 reveals the estimated effects of *Female* among Democrats and Republicans with associated confidence intervals and shows the estimated effects of *Republican* for males and females, again with associated confidence intervals. In the top panel, we see that the confidence interval for the marginal effect of *Female* among Democrats includes the value of zero whereas that among Republicans does not. This graph shows that the effect of gender among Democrats does not differ statistically distinguishably from zero but the effect of gender

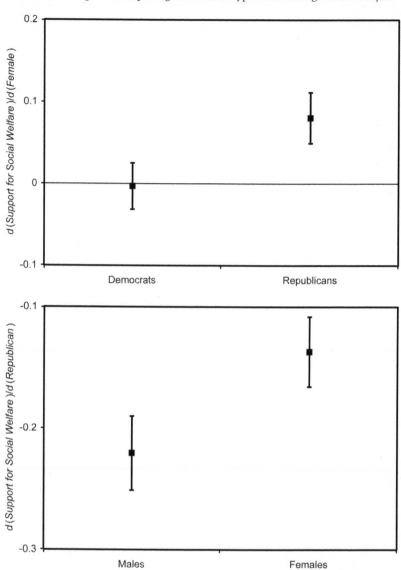

Fig. 7.   Marginal effect of *Female* and *Republican,* with 95 percent confidence intervals

among Republicans does. Furthermore, the confidence intervals do not overlap, indicating that the effect of gender differs significantly between Democrats and Republicans. In the bottom panel, zero lies outside both sets of confidence intervals; the marginal effect of partisanship is significantly different from zero for both males and females. Again, the confidence intervals do not overlap, suggesting that the marginal effect of partisanship is significantly stronger (in the negative direction) among males.

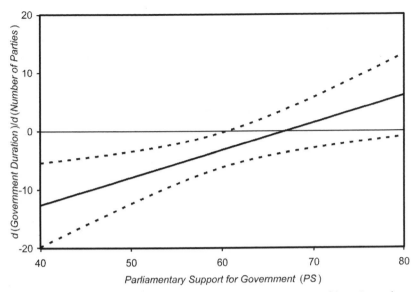

Fig. 8.   Marginal effect of *Number of Parties,* with 90 percent confidence interval

Moving to our government-durability example, figures 8 and 9 illustrate the marginal effect of the number of governing parties and the marginal effect of parliamentary support on government duration from the simple, linear-interactive model of government duration featured in table 5. Figure 8 shows that the marginal effect of *NP* takes negative and positive values, depending on the value of *PS,* as we noted in that discussion. It also reveals far more clearly than discussion alone could that, at lower values of *PS,* the (negative) marginal effect of *NP* is statistically distinguishable from zero (the 90 percent confidence interval lies entirely below zero until parliamentary support reaches about 62 percent). While the estimated effect becomes positive beyond that value, it remains statistically indistinguishable from zero through the rest of the sample range. We can conclude reasonably confidently that the number of governing parties reduces government duration for parliamentary support below 62 percent, as expected, and we could note that it merely becomes statistically indistinguishable from zero beyond that, even though estimates suggest that it might even become positive. Analogously, figure 9 plots the estimated marginal effect of parliamentary support on government duration as a function of the number of governing parties. It is generally positive and becomes statistically distinguishable from zero in that direction once the number of governing parties reaches two.

Recall that figure 2 plotted estimated government duration as a quadratic function of parliamentary support. It also plotted the estimated

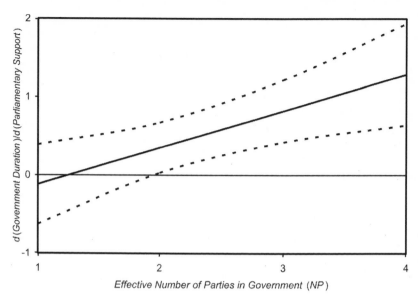

Fig. 9. Marginal effect of *Parliamentary Support for Government*, with 90 percent confidence interval

marginal effect of parliamentary support on government duration as a function of the level of support, based on the quadratic model estimated. Graphical presentation of estimates and estimated effects in nonlinear models is especially useful, and including some representation of the certainty of those estimates and estimated effects is equally crucial. Accordingly, figure 10 adds 90 percent confidence intervals to the straight line (the estimated marginal conditional effect line) in figure 2, using the square root of the expression in (27) to calculate the estimated standard error of the estimated marginal conditional effect. (We discuss construction of the confidence interval around the curved line, the predicted values, subsequently.) We take the estimated marginal effect and add (subtract) the product of the t-critical value and the estimated standard error to obtain the upper (lower) bound of the confidence interval:

$$(\hat{\beta}_{ps} + 2\hat{\beta}_{ps^2}PS) \pm 1.729 \times [\widehat{V(\hat{\beta}_{ps})} + 4PS^2 \times \widehat{V(\hat{\beta}_{ps^2})} + 4PS$$
$$\times \widehat{C(\hat{\beta}_{ps},\hat{\beta}_{ps^2})}]^{0.5}$$

Likewise, figure 3 plotted the estimated marginal nonlinear conditional effect of parliamentary support on government duration from the model specifying *PS* in natural log terms and interactively with the number of governing parties, *NP*. This presentation, too, requires indication of the uncertainty of these estimated effects. We first use the expression

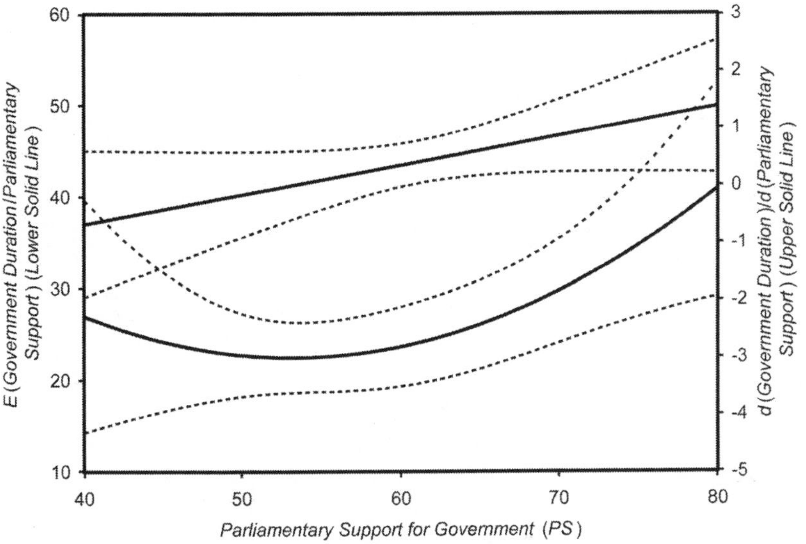

Fig. 10. Marginal effect of *Parliamentary Support* and predicted *Government Duration*, quadratic-term model, with 90 percent confidence intervals

provided in equation (28) to calculate the estimated standard error of the marginal effect of *PS* and then add (subtract) the product of the estimated standard error and the *t*-critical value to the estimated marginal effect to obtain the upper (lower) bound of the confidence interval:

$$(\hat{\beta}_{ps} + \hat{\beta}_{np\ln(ps)}NP)/PS \pm 1.74 \times \left[\frac{1}{PS^2}\left(\widehat{V(\hat{\beta}_{\ln(ps)})} + NP^2\widehat{V(\hat{\beta}_{np\ln(ps)})}\right.\right.$$
$$\left.\left. + 2NP \times \widehat{C(\hat{\beta}_{\ln(ps)},\hat{\beta}_{np\ln(ps)})}\right)\right]^{0.5}$$

To accommodate the two-dimensional, monochrome technology of most print publications and to reduce visual clutter, figure 11 plots just two of these conditional-effect lines with confidence intervals, those corresponding to the revealing and interesting $NP = 2$ and $NP = 4$ cases.

The estimated marginal conditional effects of the number of governing parties on government duration can also be plotted along values of parliamentary support for government, with a confidence interval. We calculate the confidence interval as

$$(\hat{\beta}_{np} + \hat{\beta}_{np\ln(ps)}\ln(PS)) \pm 1.74 \times \left[\widehat{V(\hat{\beta}_{np})} + (\ln(PS))^2 \times \widehat{V(\hat{\beta}_{np\ln(ps)})}\right.$$
$$\left. + 2\ln(PS) \times \widehat{C(\hat{\beta}_{np},\hat{\beta}_{np\ln(ps)})}\right]^{0.5}$$

As figure 12 reveals, the point estimate of the effect of *NP* does turn positive beyond $PS \approx 65$ percent. However, this putatively positive effect

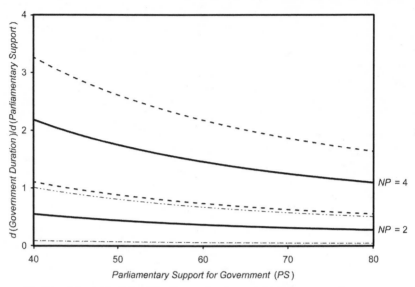

Fig. 11.   Marginal effect of *Parliamentary Support for Government,* log-transformation interactive model, with 90 percent confidence intervals

never surpasses even generous levels of statistical significance ($p < 0.10$), whereas the decreasingly negative effects in the range below $PS \approx 60$ percent are statistically distinguishable from zero at this level. Thus, this fuller picture of the evidence from the empirical analysis rather suggests that, as expected intuitively, increasing government fractionalization reduces durability, but this detrimental effect generally diminishes as the strength of parliamentary support for that fractionalized government rises.

As we saw comparing the regression output from the table 5 (linear-interactive) and table 8 (log-transformed-interactive) versions of this model, the curvature of the effect lines induced by the log-transformation of *PS* is not especially strongly supported relative to a linear specification ($\bar{R}^2 = 0.520$ vs. $\bar{R}^2 = 0.511$). Graphically, this relatively weak support is seen from how easily straight conditional-effect lines could fit within the confidence intervals surrounding these slightly curved conditional-effect lines. However, we caution that exact correspondence to the significance with which the non-linear-interactive could reject the linear-interactive model does not emerge from these graphs. In fact, more generally, ability to draw flat (unconditional) effect lines within the confidence intervals of slanted (conditional) effect lines does not correspond to a hypothesis test that the effect is conditional (interactive). The correct test of that, as table 8 detailed, is the simple *t*-test of the interaction term in the

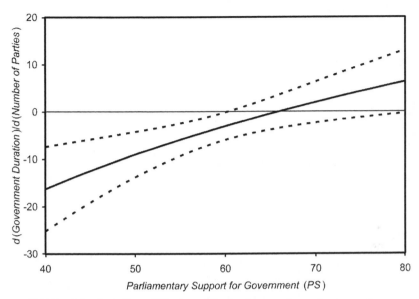

Fig. 12. Marginal effect of *Number of Parties*, log-transformation interactive model, with 90 percent confidence interval

model (or analogous *F*-tests in multiple-interaction models as in table 9). Significance of the hypothesis that the effect of *x* depends on *z* (i.e., generally) does not guarantee that the confidence intervals for the conditional effects at the high and low end of the range of *z* plotted or, for that matter, necessarily at any two *z*-values plotted, will fail to overlap.[46]

In the chained three-way-interaction model of the first column of table 9, the effects of *PS* and of party discipline, *PD*, are conditioned by one other variable, *NP*. We have already discussed and demonstrated how to present this type of conditional effect. Note, though, that the effect of *NP* in this model depends on not one but two other variables: *PS* and *PD*: $\partial \widehat{GD}/\partial NP = \hat{\beta}_{np} + \hat{\beta}_{npps}PS + \hat{\beta}_{nppd}PD$. One might consider a three-dimensional plot of such a conditional effect, plotting the marginal effect of *NP* (*y*-axis) as a function of *PS* (*x*-axis) and of *PD* (*z*-axis). However, conditional-effect "lines" in such cases will actually be planes plotted at linearly changing heights *y* as *x* and *z* change, which would be difficult to render clearly on two-dimensional pages, especially since we must also include confidence intervals, which will be (hourglass) curved

---

46. Indeed, in this case, the linear and the nonlinear models are nonnested and have the same degrees of freedom, and so empirical comparison of the linear versus nonlinear models must proceed on other bases entirely.

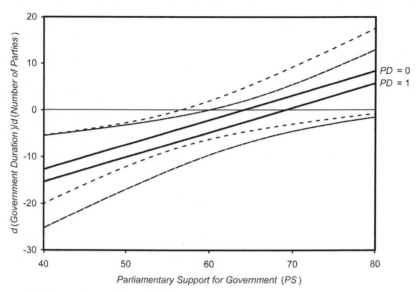

Fig. 13. Marginal effect of *Number of Parties,* chained-interaction model, with 90 percent confidence intervals

surfaces above and below that conditional-effect plane. We therefore rec-ommend eschewing three-dimensional graphics and instead plotting contours of those three-dimensional relationships onto two dimensions. To be precise, we suggest plotting $\partial \hat{y}/\partial x = \hat{\beta}_x + \hat{\beta}_{xz}z + \hat{\beta}_{xw}w$ as a function of z or w at a few values of w or z, each of which will generate one conditional-effect line, each with its own confidence interval, like those previously shown. In this case, PD is binary, so we could plot $\widehat{\partial GD/\partial NP} = \hat{\beta}_{np} + \hat{\beta}_{npps}PS + \hat{\beta}_{nppd}PD$ as a function of PS just at $PD = 0$ and at $PD = 1$, with confidence intervals, to illustrate the estimated conditional effects fully. Figure 13 demonstrates that the detrimental effect of NP on government durability declines with PS, but it does not seem to be further conditioned by PD in this analysis.

In the pairwise and fully interactive three-way-interaction models, finally, the effects of NP, PS, and PD each depend on the other two factors. Figures 14 and 15 demonstrate how researchers can graph estimation results from pairwise-interaction models effectively. Figure 14 parallels the case of figure 13, plotting how the effect of parliamentary support depends on the number of governing parties and party discipline. (The effect of NP symmetrically depends on PS and PD in this model, too, but those results and that figure add little to what fig. 13 already displayed.) The formulas for the effect lines in this figure parallel those from before also:

$$\widehat{\partial GD/\partial PS} = \hat{\beta}_{ps} + \hat{\beta}_{npps}NP + \hat{\beta}_{pspd}PD$$

$$\widehat{V(\partial GD/\partial PS)} = \widehat{V(\hat{\beta}_{ps})} + NP^2 \times \widehat{V(\hat{\beta}_{npps})}$$

$$+ PD^2 \times \widehat{V(\hat{\beta}_{pspd})} + 2NP \times \widehat{C(\hat{\beta}_{ps},\hat{\beta}_{npps})}$$

$$+ 2PD \times \widehat{C(\hat{\beta}_{ps},\hat{\beta}_{pspd})} + 2NP \times PD \times \widehat{C(\hat{\beta}_{npps},\hat{\beta}_{pspd})}$$

$$90\% \text{ c.i.: } \frac{\widehat{\partial GD}}{\partial PS} \pm 1.75 \times \left[\widehat{V\left(\frac{\partial GD}{\partial PS}\right)} \right]^{0.5}$$

The nearly nonoverlapping confidence intervals in figure 14 reveal that the effect of parliamentary support, unlike that of the number of governing parties (not shown), does seem to depend somewhat on party discipline. Intuitively, the durability-enhancing effects of larger parliamentary support are greater with higher than with lower discipline of those additional partisan supporters. The upward slopes of these conditional-effect lines show also that the benefit of greater parliamentary support to government durability seems to increase with the fractionalization of those governments. Intuitively, single-party governments can survive with bare-majority support; multiparty governments need more cushion. This feature also seems more statistically certain at higher party discipline, as the narrower confidence region for the effect at $PD = 1$ than at $PD = 0$ reveals. The effect of party discipline in these models depends on two continuous variables, NP and PS.

Therefore, three dimensions are needed to represent its conditional effects fully; however, a pair of two-dimensional graphs can suffice nearly as fully and will usually be far easier to comprehend. Namely, we recommend plotting $\widehat{\partial GD/\partial PD} = \hat{\beta}_{pd} + \hat{\beta}_{nppd}NP + \hat{\beta}_{pspd}PS$ as a function of NP at a few values of PS and as a function of PS at a few values of NP, each with confidence intervals as in figure 15.[47] The upper graph displays two flat conditional-effect lines and nearly completely nonoverlapping confidence intervals. The lower graph displays two clearly upward-sloping conditional-effect lines nearly on top of each other and with almost fully overlapping confidence intervals. These graphs suggest that the effect of

---

47. Using the now-familiar procedures, the estimated variance of the effect is calculated as

$$\widehat{V\left(\frac{\partial GD}{\partial PD}\right)} = \widehat{V(\hat{\beta}_{pd})} + NP^2 \times \widehat{V(\hat{\beta}_{nppd})} + PS^2 \times \widehat{V(\hat{\beta}_{pspd})}$$

$$+ 2NP \times \widehat{C(\hat{\beta}_{pd},\hat{\beta}_{nppd})} + 2PS \times \widehat{C(\hat{\beta}_{pd},\hat{\beta}_{pspd})} + 2NP \times PS \times \widehat{C(\hat{\beta}_{nppd},\hat{\beta}_{pspd})}$$

Accordingly, the confidence interval is the estimated effect from the text plus or minus the t critical value times the square root of this expression.

party discipline on government duration seems to depend on parliamentary support but not on the number of governing parties in this model.

Figures 16 and 17 graph the estimated marginal effects of *NP* and *PD* in the fully interactive model wherein the effect of each variable depends on the values and the combination of the values of the other two variables.[48] Effective graphing techniques for fully interactive models mirror those for pairwise-interaction models because in both cases the effect of each variable depends on two others. The difference here is that when, as in figure 16, for example, plotting the marginal effect of one variable, for instance, *PS*, as a function of a second, *NP*, at different values of the third, *PD*, the marginal-effect lines will not be parallel because the effect of the first depends not just additively on the other two

---

48. The expressions for the estimated marginal effect of one variable in a generic three-way fully interactive model and the estimated variance of that estimated effect are

$$\partial \hat{y}/\partial x = \hat{\beta}_x + \hat{\beta}_{xz}z + \hat{\beta}_{xw}w + \hat{\beta}_{xzw}zw$$

$$V\left(\widehat{\frac{\partial \hat{y}}{\partial x}}\right) = \widehat{V(\hat{\beta}_x)} + z^2\widehat{V(\hat{\beta}_{xz})} + w^2\widehat{V(\hat{\beta}_{xw})} + z^2w^2\widehat{V(\hat{\beta}_{xzw})}$$

$$+ 2z\widehat{C(\hat{\beta}_x,\hat{\beta}_{xz})} + 2w\widehat{C(\hat{\beta}_x,\hat{\beta}_{xw})} + 2zw\widehat{C(\hat{\beta}_x,\hat{\beta}_{xzw})}$$

$$+ 2zw\widehat{C(\hat{\beta}_{xz},\hat{\beta}_{xw})} + 2z^2w\widehat{C(\hat{\beta}_{xz},\hat{\beta}_{xzw})} + 2zw^2\widehat{C(\hat{\beta}_{xw},\hat{\beta}_{xzw})}$$

More simply, in matrix notation: $\partial \hat{y}/\partial x = \mathbf{m}'\hat{\boldsymbol{\beta}} = \begin{bmatrix} 1 & z & w & zw \end{bmatrix} \begin{bmatrix} \hat{\beta}_x \\ \hat{\beta}_{xz} \\ \hat{\beta}_{xw} \\ \hat{\beta}_{xzw} \end{bmatrix}$

$$V(\mathbf{m}'\hat{\boldsymbol{\beta}}) = \mathbf{m}'V(\hat{\boldsymbol{\beta}})\mathbf{m} = \begin{bmatrix} 1 & z & w & zw \end{bmatrix} \times V \begin{bmatrix} \hat{\beta}_x \\ \hat{\beta}_{xz} \\ \hat{\beta}_{xw} \\ \hat{\beta}_{xzw} \end{bmatrix} \times \begin{bmatrix} 1 \\ z \\ w \\ zw \end{bmatrix}$$

$$= \begin{bmatrix} 1 & z & w & zw \end{bmatrix} \begin{bmatrix} V(\hat{\beta}_x) & C(\hat{\beta}_x,\hat{\beta}_{xz}) & C(\hat{\beta}_x,\hat{\beta}_{xw}) & C(\hat{\beta}_x,\hat{\beta}_{xzw}) \\ C(\hat{\beta}_x,\hat{\beta}_{xz}) & V(\hat{\beta}_{xz}) & C(\hat{\beta}_{xz},\hat{\beta}_{xw}) & C(\hat{\beta}_{xz},\hat{\beta}_{xzw}) \\ C(\hat{\beta}_x,\hat{\beta}_{xw}) & C(\hat{\beta}_{xz},\hat{\beta}_{xw}) & V(\hat{\beta}_{xw}) & C(\hat{\beta}_{xw},\hat{\beta}_{xzw}) \\ C(\hat{\beta}_x,\hat{\beta}_{xzw}) & C(\hat{\beta}_{xz},\hat{\beta}_{xzw}) & C(\hat{\beta}_{xw},\hat{\beta}_{xzw}) & V(\hat{\beta}_{xzw}) \end{bmatrix} \begin{bmatrix} 1 \\ z \\ w \\ zw \end{bmatrix}$$

In words, the variance of a sum of random variables and constants, such as an estimated conditional effect, is the sum of all the variances of the variables (the estimated coefficients implied by the conditional effect), each multiplied by the square of their cofactor (the associated independent variable(s)), plus two times each of the covariances of the variables (the estimated coefficients) times the product of their cofactors (the associated independent variable(s)).

To complete the set of graphs, the marginal effect of *PS* can also be graphed following similar procedures.

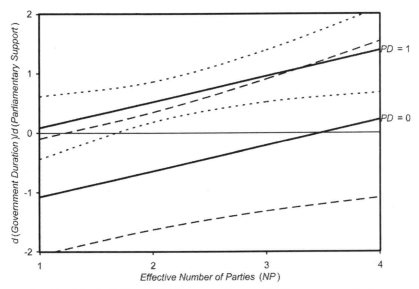

Fig. 14. Marginal effect of *Parliamentary Support for Government,* pairwise-interaction model, with 90 percent confidence intervals

but multiplicatively as well. The other major difference is the magnitude of the confidence intervals; attempting to estimate such complexly interactive relations, with seven nested, and so highly colinear, linear-interaction terms, with just twenty-two observations and fourteen degrees of freedom, will almost always prove quixotic, as it does here. We can distinguish from zero even at the low $p = 0.10$ level only (1) the intuitive increasingly beneficial effect of parliamentary support as the number of parties increases in a high party-discipline environment (fig. 16, $PD = 1$ line), (2) the converse increasingly beneficial effect of party discipline as parliamentary support for a government of relatively few parties surpasses about 50 percent (fig. 17b, $NP = 2$ line), and (3) the decreasingly beneficial effect of party discipline as the number of parties in a high parliamentary-support government rises (fig. 17a, $PS = 80$ line). Almost none of these estimated complexly conditional marginal effects is distinguishable from any other at almost any combination of independent-variable values. Researchers interested in exploring such complex context-conditionality empirically face challenges. This example illustrates the importance of maximizing observations and degrees of freedom and of leveraging theory to specify interactive hypotheses as precisely as possible (as strongly urged in chap. 2 and as demonstrated in Franzese 1999, 2002, 2003a).

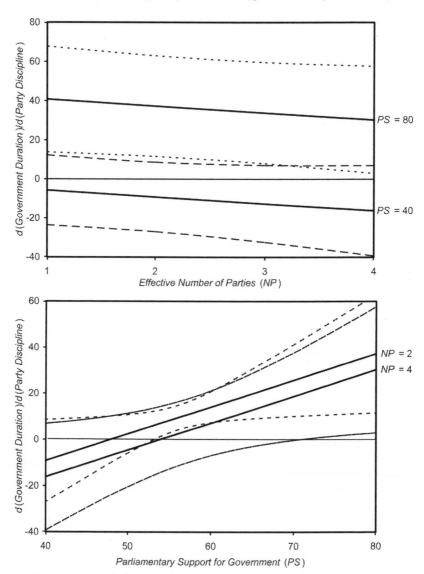

Fig. 15. Marginal effect of *Party Discipline,* pairwise-interaction model, with 90 percent confidence intervals

Presentation of Predicted Values

Aside from presenting conditional effects, researchers may also wish to present the predictions of y as x varies across a range of values, say, from x_a to x_c, its sample minimum to maximum, while holding z constant at some (meaningful and revealing) value. Changes in these predictions from

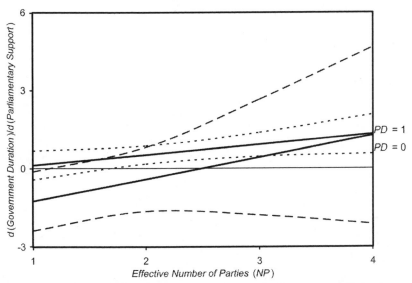

Fig. 16. Marginal effect of *Parliamentary Support for Government,* fully interactive model, with 90 percent confidence intervals

some particular $\hat{y}|x_a$ to $\hat{y}|x_c$ would reveal the effects of such changes in x on y at that level of z as just discussed, but we may also wish to present tables or graphs of predictions per se as x varies, holding z fixed. (Recall that xz will also vary with x, even though z is held constant.) Including measures of uncertainty around these predictions is again imperative, and, as with effects, each predicted value at some particular x and z values has its own level of uncertainty attached to it. Thus, tables and graphs of predicted values should also include standard errors and/or confidence intervals (variances, standard errors, significance levels) around each of those predicted values.

In the standard linear-interaction model, the variance around each predicted value is

$$V(\hat{y} \mid x,z) = V(\hat{\gamma}_0 + \hat{\beta}_x x + \hat{\beta}_z z + \hat{\beta}_{xz} xz) \tag{29}$$

Expanding this expression:[49]

$$\begin{aligned} V(\hat{y}) = {} & V(\hat{\gamma}_0) + x^2 V(\hat{\beta}_x) + z^2 V(\hat{\beta}_z) + (xz)^2 V(\hat{\beta}_{xz}) \\ & + 2x C(\hat{\gamma}_0, \hat{\beta}_x) + 2z C(\hat{\gamma}_0, \hat{\beta}_z) + 2xz C(\hat{\gamma}_0, \hat{\beta}_{xz}) \\ & + 2xz C(\hat{\beta}_x, \hat{\beta}_z) + 2x(xz) C(\hat{\beta}_x, \hat{\beta}_{xz}) + 2z(xz) C(\hat{\beta}_z, \hat{\beta}_{xz}) \end{aligned} \tag{30}$$

---

49. Note 30 gives the more general linear-algebraic formula for variances of linear combinations of random variables and constants.

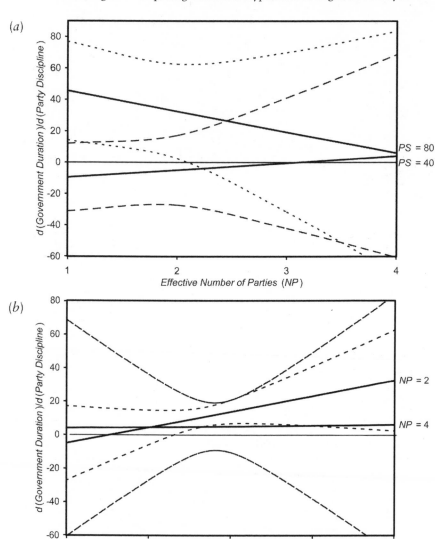

Fig. 17. Marginal effect of *Party Discipline,* fully interactive model, with 90 percent confidence intervals

In words, the variance of a sum equals the sum of the variances plus two times all the covariances. More completely, the variance of a sum of random variables (here, the coefficient estimates) times constants (here, independent variables) is equal to the sum of the variances times the associated constants squared plus two times all the covariances times the product of their constant cofactors.[50] As before, we will need the estimated variance-covariance matrix of the parameter estimates $(\mathbf{V}(\hat{\boldsymbol{\beta}}))$ to calculate this, which can be easily recalled by an additional postestimation command in most statistical software.[51]

Let us use our first empirical example to calculate the predicted number of presidential candidates corresponding with various values of *Ethnic Groups* and *Runoff* along with the variance of each predicted value. Table 13 gave the variance-covariance matrix of the coefficient estimates from this model. When *Groups* = 1 and *Runoff* = 0, we predict the number of candidates to be

$$(\widehat{Candidates} \mid Groups = 1, Runoff = 0) = 4.303 - 0.979 \times 1$$

$$- 2.491 \times 0 + 2.005$$

$$\times 1 \times 0 = 3.324$$

Using equation (30), substituting *Groups* = 1 and *Runoff* = 0, yields the following expression:

$$V(\widehat{Candidates} \mid Groups = 1, Runoff = 0) = \widehat{V(\hat{\beta}_0)} + 1^2 \widehat{V(\hat{\beta}_G)}$$

$$+ 0^2 \widehat{V(\hat{\beta}_R)} + (0)(1)^2 \ \widehat{V(\hat{\beta}_{GR})}$$

$$+ 2 \times 1 \times \widehat{C(\hat{\beta}_0,\hat{\beta}_G)}$$

$$+ 2 \times 0 \times \widehat{C(\hat{\beta}_0,\hat{\beta}_R)}$$

$$+ 2 \times 1 \times 0 \times \widehat{C(\hat{\beta}_0,\hat{\beta}_{GR})}$$

$$+ 2 \times 1 \times 0 \times \widehat{C(\hat{\beta}_G,\hat{\beta}_R)}$$

$$+ 2 \times 1 \times (1 \times 0) \ \widehat{C(\hat{\beta}_G,\hat{\beta}_{GR})}$$

$$+ 2 \times 0 \times (1 \times 0) \ \widehat{C(\hat{\beta}_R,\hat{\beta}_{GR})}$$

Substituting the estimated values of the variances and covariances of the coefficients:

$$V(\widehat{Candidates} \mid Groups = 1, Runoff = 0) = 1.509 + 0.593 + 2$$

$$\times (-0.900) = 0.302$$

---

50. These are variances and confidence intervals for $E(y|x,z = z_0)$ and not forecast or prediction errors, which would include also some uncertainty due to the variance of the regression's error term. See note 30.

51. Appendix B provides step-by-step STATA commands.

Table 22 presents the standard errors of each of the predicted values as *Ethnic Groups* ranges from one to three, when *Runoff* takes the values of zero and one. These predictions can also be graphed as described later.

Obviously, these calculations will become quite cumbersome, quite quickly, in the presence of additional covariates. In fact, calculation of the variance of predicted values requires attention to the levels of all the independent variables and to the variance of each estimated coefficient and the covariances between each of the estimated coefficients. In our simple model, which includes just three variables plus an intercept, this involves ten terms. Adding just one more regressor (which did not interact with any others) would require us to include five more terms in equation (30)!

One way to simplify the expression is to use matrix algebra to depict $\hat{y}$ and to calculate $\widehat{V(\hat{y})}$ (see note 30). Note that a predicted value, $\hat{y}$, sums the products of sets of values of the right-hand-side variables and their corresponding coefficients. Let $\mathbf{M_h}$ be a *j*-by-*k* matrix of values at which x, z, and any other variables of interest in the equation are set, where *j* refers to the number of values at which the predicted value is calculated and *k* refers to the number of regressors, including the constant. Suppose we were to hold z (and any of the other variables) at some logically relevant value(s), say, z_0, and examine the predicted values of $\hat{y}$ at a set of *j* evenly spaced values of x from x_a to x_c and correspondingly, as xz takes *j* evenly spaced values from $x_a z_0$ to $x_c z_0$. In our standard equation, we have estimated coefficients for x, z, and xz, in addition to an intercept. Matrix $\mathbf{M_h}$ is thus

$$\mathbf{M_h} = \begin{bmatrix} x_a & z_0 & x_a z_0 & 1 \\ x_{a+1} & z_0 & x_{a+1} z_0 & 1 \\ \vdots & \vdots & \vdots & 1 \\ x_c & z_0 & x_c z_0 & 1 \end{bmatrix}$$

In $\mathbf{M_h}$, the value of x increments evenly from some value x_a to some other value x_c; z is fixed at z_0; and the interaction term xz varies as x does. The

TABLE 22.   Confidence Intervals for Predicted *Number of Presidential Candidates*

	\multicolumn{3}{c}{Runoff = 0}			\multicolumn{3}{c}{Runoff = 1}		
	$\hat{y}$	s.e.($\hat{y}$)	90% Confidence Interval	$\hat{y}$	s.e.($\hat{y}$)	90% Confidence Interval
Groups = 1	3.324	0.550	[2.344, 4.305]	2.838	0.512	[1.925, 3.752]
Groups = 1.5	2.835	0.380	[2.158, 3.511]	3.351	0.387	[2.662, 4.041]
Groups = 2	2.345	0.532	[1.397, 3.292]	3.865	0.437	[3.104, 4.625]
Groups = 2.5	1.855	0.847	[0.345, 3.365]	4.378	0.600	[3.308, 5.447]
Groups = 3	1.366	1.204	[−0.780, 3.512]	4.891	0.827	[3.417, 6.364]

column of ones represents the constant (intercept). We can then express the vector of predicted values $\hat{y}$ as

$$\hat{y} = M_h\hat{\beta} \quad \text{where } \hat{\beta} = \begin{bmatrix} \hat{\beta}_x \\ \hat{\beta}_z \\ \hat{\beta}_{xz} \\ \hat{\beta}_0 \end{bmatrix}$$

As a consequence, $V(\hat{y}) = V(M_h\hat{\beta})$. Since M_h is a matrix of values at which we set our independent variables, and since independent variables are fixed in repeated sampling under classical regression assumptions, the matrix M_h is a constant whereas $\hat{\beta}$ is a random vector. Accordingly

$$V(\hat{y}) = V(M_h\hat{\beta}) = M_hV(\hat{\beta})M_h'$$

where $V(\hat{\beta})$ is the variance-covariance matrix of the estimated coefficients.

The j diagonal elements in $V(\hat{y})$ correspond with the variances of the j predicted values of $\hat{y}$ at various values included in M_h. As before, we denote the estimate of $V(\hat{\beta})$ as $\widehat{V(\hat{\beta})}$.

Using our *Candidates* example, we can calculate the variance of the predicted values of y as follows. First, varying values of *Groups* in 0.5 intervals from 1 to 3, holding *Runoff* to 0, gives

$$M_h = \begin{bmatrix} 1 & 0 & 0 & 1 \\ 1.5 & 0 & 0 & 1 \\ 2 & 0 & 0 & 1 \\ 2.5 & 0 & 0 & 1 \\ 3 & 0 & 0 & 1 \end{bmatrix}$$

The first column indicates the values of *Groups*, the second column indicates the values of *Runoff*, the third column indicates the values of *Groups* $\times$ *Runoff*, and the fourth column represents the values for the intercept. The estimated variances of the predicted numbers of candidates at these values are therefore given by

$$\widehat{V(\hat{y})} = M_h\widehat{V(\hat{\beta})}M_h' = \begin{bmatrix} 1 & 0 & 0 & 1 \\ 1.5 & 0 & 0 & 1 \\ 2 & 0 & 0 & 1 \\ 2.5 & 0 & 0 & 1 \\ 3 & 0 & 0 & 1 \end{bmatrix} \begin{bmatrix} 0.593 & 0.900 & -0.593 & -0.900 \\ 0.900 & 2.435 & -1.377 & -1.509 \\ -0.593 & -1.377 & 0.885 & 0.900 \\ -0.900 & -1.509 & 0.900 & 1.509 \end{bmatrix}$$

$$\times \begin{bmatrix} 1 & 1.5 & 2 & 2.5 & 3 \\ 0 & 0 & 0 & 0 & 0 \\ 0 & 0 & 0 & 0 & 0 \\ 1 & 1 & 1 & 1 & 1 \end{bmatrix}$$

which produces the following symmetric matrix:

$$\widehat{V(\hat{y})} = \begin{bmatrix} 0.302 & 0.149 & -0.005 & -0.159 & -0.312 \\ 0.149 & 0.143 & 0.138 & 0.133 & 0.128 \\ -0.005 & 0.138 & 0.281 & 0.424 & 0.567 \\ -0.159 & 0.133 & 0.424 & 0.715 & 1.007 \\ -0.312 & 0.128 & 0.567 & 1.007 & 1.446 \end{bmatrix}$$

The diagonal elements are $\widehat{V(\hat{y})}$ for the respective values of *Groups* when *Runoff* = 0. Statistical software or a basic spreadsheet program can make these matrix calculations simple to implement.

Predicted values are often more effectively displayed when graphed with confidence intervals, which can be constructed as $\hat{y} \pm t_{df,p} \sqrt{\widehat{V(\hat{y})}}$, where, as before, $t_{df,p}$ is the critical value in a t-distribution with df degrees of freedom that produces a p-value corresponding to half of the probability outside of the desired confidence interval. For example, lower and upper bounds of a 95 percent confidence interval will again come from $t_{df,p}$ of approximately 1.96 in large samples.

For this example, we calculate $\hat{y}$ along evenly spaced values of *Groups* from one to three, fixing *Runoff* first to zero and then to one. To calculate confidence intervals, we need to calculate the variances of these predicted values and to identify a desired level of confidence. Given our small sample, we again accept appreciable uncertainty, selecting a 90 percent confidence interval, implying a critical value of $t_{12,\alpha=0.10} = 1.782$. The upper bound and lower bound for the confidence intervals are therefore

Upper bound: $\hat{y} + 1.782 \times \sqrt{\widehat{V(\hat{y})}}$

Lower bound: $\hat{y} - 1.782 \times \sqrt{\widehat{V(\hat{y})}}$

For *Groups* = 1 and *Runoff* = 0, for example, $\widehat{V(\hat{y})} = 0.302$ as seen earlier, and so the 90 percent confidence interval is

Upper bound: $3.324 + 1.782 \times 0.302 = 4.304$

Lower bound: $3.324 - 1.782 \times 0.302 = 2.345$

Table 22 displays the confidence intervals calculated for the predicted values of the number of presidential candidates as *Groups* ranges from 1 to 3 in steps of 0.5, with *Runoff* fixed to 0 and to 1. Figure 18 graphs these predicted values and confidence intervals with *Groups* on the x-axis, the predicted values on the y-axis, and the value of *Runoff* fixed.

Figure 18 displays straight lines, indicating how the predicted number of candidates changes as *Groups* varies, in the presence and absence

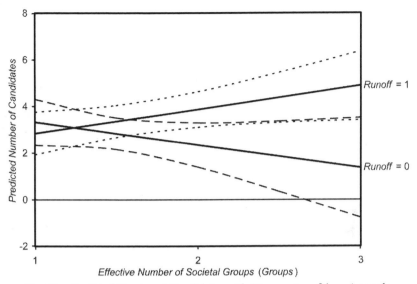

Fig. 18.   *Predicted Number of Candidates,* with 90 percent confidence intervals

of a runoff system. The hourglass curves indicate the degree of certainty associated with each predicted value, $\hat{y}$. As with estimated effects, these predictions have greatest certainty around the mean of *Groups* and less certainty at more extreme and, especially, out-of-sample values.

Table 22 reinforces what we have already seen in this example: substantial overlap in the 90 percent confidence intervals for the predicted number of presidential candidates in the presence and absence of a runoff system when only one ethnic group exists but much less overlap in these confidence intervals at higher numbers of *Groups*. These results suggest that the impact of runoff systems on the number of candidates becomes more discernibly positive statistically as *Groups* increases.

Tables 23 and 24 provide confidence intervals for various predicted values in two of our other examples: the U.S. support for social welfare model and the baseline model of government duration (with just the interaction between the number of governing parties and parliamentary support). The results in table 23 are easily comprehensible, given that the interaction involves only two binary variables: *Female* and *Republican*. Table 23 shows the negligible difference in predicted social-welfare support among Democrats (*Republican* = 0) by gender, with the confidence intervals around those predicted values overlapping substantially (in fact, the confidence interval for male Democrats entirely encloses the confidence interval for female Democrats). We also see that social-welfare support among Republican males is appreciably lower than among Republican females. This gender gap among Republicans is statistically distinguishable from zero in that the confidence intervals for male and female Republicans do not overlap. The same information could also be presented graphically, but the simplicity of the table may recommend tabular form instead.

Table 24 adds 90 percent confidence intervals to the predicted values presented in table 6. It is less immediately interpretable, given the plethora of values that NP and PS can take. A graph may be the most effective means of presenting the predicted values and their associated confidence intervals in cases like this, as shown in figure 19.

For variables that enter nonlinearly—for instance, in the example where parliamentary support for government is quadratically related to government durability—the procedure previously outlined still obtains. Recall that figure 2 plotted estimated government duration as a quadratic function of parliamentary support. Accordingly, figure 10 adds 90 percent confidence intervals to the predicted value curve, using the results from the model in table 7 and calculating the confidence interval by

$$\widehat{GD} = \hat{\beta}_0 + \hat{\beta}_{ps}PS + \hat{\beta}_{ps2}PS^2$$

$$\widehat{V(\widehat{GD})} = \widehat{V(\hat{\beta}_0 + \hat{\beta}_{ps}PS + \hat{\beta}_{ps2}PS^2)}$$
$$= \widehat{V(\hat{\beta}_0)} + PS^2 \times \widehat{V(\hat{\beta}_{ps})} + PS^4\widehat{V(\hat{\beta}_{ps2})} + 2PS \times \widehat{C(\hat{\beta}_0,\hat{\beta}_{ps})}$$
$$+ 2PS^2 \times \widehat{C(\hat{\beta}_0,\hat{\beta}_{ps2})} + 2PS^3 \times \widehat{C(\hat{\beta}_{ps},\hat{\beta}_{ps2})}$$

$$90\% \text{ c.i.} = (\hat{\beta}_0 + \hat{\beta}_{ps}PS + \hat{\beta}_{ps2}PS^2)$$
$$\pm 1.73 \left[\begin{array}{l} \widehat{V(\hat{\beta}_0)} + PS^2 \times \widehat{V(\hat{\beta}_{ps})} + PS^4\widehat{V(\hat{\beta}_{ps2})} \\ + 2PS \times \widehat{C(\hat{\beta}_0,\hat{\beta}_{ps})} + 2PS^2 \times \widehat{C(\hat{\beta}_0,\hat{\beta}_{ps2})} \\ + 2PS^3 \times \widehat{C(\hat{\beta}_{ps},\hat{\beta}_{ps2})} \end{array} \right]^{0.5}$$

In the log-transformed-PS model, we can also graph the predicted government duration, at selected values of NP and PD, along values of

TABLE 23.   Confidence Intervals for Predicted *Support for Social Welfare*

	Republican = 0			Republican = 1		
	$\hat{y}$	$s.e.(\hat{y})$	95% Confidence Interval	$\hat{y}$	$s.e.(\hat{y})$	95% Confidence Interval
Female = 0	0.745	0.0110	[0.724, 0.767]	0.525	0.0110	[0.503, 0.546]
Female = 1	0.742	0.0094	[0.724, 0.760]	0.605	0.0113	[0.583, 0.627]

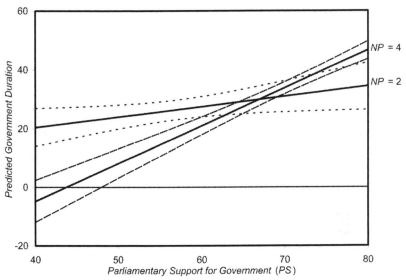

Fig. 19. Predicted *Government Duration,* with 90 percent confidence intervals

parliamentary support. The 90 percent confidence interval can be calculated around $\widehat{GD}$ as

$$\widehat{GD} = \hat{\beta}_0 + \hat{\beta}_{np}NP + \hat{\beta}_{\ln(ps)}\ln(PS) + \hat{\beta}_{np\ln(ps)}NP \times \ln(PS)$$
$$+ \hat{\beta}_{pd}PD$$

$$\widehat{V(GD)} = \overline{V(\hat{\beta}_0 + \hat{\beta}_{np}NP + \hat{\beta}_{\ln(ps)}\ln(PS) + \hat{\beta}_{np\ln(ps)}NP \times \ln(PS)}$$
$$+ \hat{\beta}_{pd}PD)$$

$$90\% \text{ c.i.} = \widehat{GD} \pm 1.74 \sqrt{\widehat{V(GD)}}$$

The estimated government duration, calculated for $NP = 2$ and $NP = 4$, when $PD = 1$, and accompanying confidence intervals appear in figure 20.

TABLE 24. Confidence Intervals for Predicted *Government Duration*

		$NP = 1$		$NP = 2$		$NP = 3$		$NP = 4$
	$\hat{y}$	90% Confidence Interval	$\hat{y}$	90% Confidence Interval	$\hat{y}$	90% Confidence Interval	$\hat{y}$	90% Confidence Interval
$PS = 40$	33.05	[23.87, 42.23]	20.42	[13.94, 26.90]	7.79	[−2.37, 17.96]	−4.84	[−21.21, 11.54]
$PS = 50$	31.87	[26.59, 37.15]	23.93	[19.86, 28.00]	15.99	[9.28, 22.69]	8.05	[−2.58, 18.67]
$PS = 60$	30.70	[25.87, 35.53]	27.44	[23.99, 30.89]	24.18	[19.92, 28.45]	20.93	[14.43, 27.43]
$PS = 70$	29.52	[21.11, 37.93]	30.95	[25.67, 36.23]	32.38	[27.58, 37.17]	33.81	[26.33, 41.29]
$PS = 80$	28.34	[15.31, 41.38]	34.46	[26.42, 42.50]	40.57	[32.86, 48.28]	46.69	[34.27, 59.10]

Note: Predicted values are calculated at given values, setting $PD = 1$.

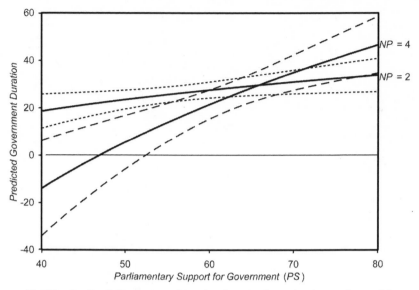

Fig. 20.   Predicted *Government Duration,* log-transformation interactive model, with 90 percent confidence intervals

Presentation of Differences of Predicted Values

Predicted values display how variation along some range of an independent variable, x, affects the level of the dependent variable, conditional upon a third independent variable, z. Researchers may sometimes wish to present the estimated effects of discrete changes rather than marginal changes of independent variables involved in interaction terms: $\Delta y/\Delta x$ rather than $\partial y/\partial x$. For example, one might want to plot the estimated effect of some substantively motivated counterfactual increase or decrease in an independent variable, say, of a 10 percent increase in parliamentary support; or of a unit change in binary indicators like gender, partisanship, runoff, or party discipline; or of a change from the level of some well-known exemplar to another (e.g., from the average number of governing parties in the United Kingdom, 1, to that of the Netherlands, 3.3). Provided that the variables involved in the estimated conditional effect enter only linearly, as in all of our examples except those using the square or natural log of parliamentary support (tables 7 and 8), doing so requires only a very simple extension of our preceding discussion of presenting marginal effects.

In regression models where the independent variables enter only linearly or linear interactively, the estimated marginal effect of any variable is equal to the estimated effect of a unit increase in that variable. In a linear-interaction model involving x, z, and xz, for example, $\partial \hat{y}/\partial x = \hat{\beta}_x +$ $\hat{\beta}_{xz}z$ and $\Delta \hat{y}/\Delta x = \hat{\beta}_x \Delta x + \hat{\beta}_{xz} (\Delta x) z$, which gives $\Delta \hat{y}/\Delta x = \hat{\beta}_x + \hat{\beta}_{xz}z$ for $\Delta x = 1$. Of course, their estimated standard errors are also identical. Thus, figures 4–9 and 13–17 all give the estimated effects of a unit increase as well as the estimated slope (or effect of a marginal increase) of their respective independent variables in their respective models.[52] More generally, if we wanted to present the effect of some discrete change other than $\Delta x = 1$ in a linear-interaction model, we need only replace the marginal effect, $\partial \hat{y}/\partial x = \hat{\beta}_x + \hat{\beta}_{xz}z$, with that of the change, $\Delta \hat{y}/\Delta x = \hat{\beta}_x \Delta x$ $+\hat{\beta}_{xz} (\Delta x)z$, which amounts simply to multiplying the marginal effect by Δx: $\Delta \hat{y}/\Delta x = \Delta x(\hat{\beta}_x + \hat{\beta}_{xz}z)$. To estimate standard errors for confidence intervals around differences in predicted values, we apply the usual variance formula: $\widehat{V(\Delta \hat{y}/\Delta x)} = (\Delta x)^2 \widehat{V(\hat{\beta}_x)} + z^2 (\Delta x)^2 \widehat{V(\hat{\beta}_{xz})} + 2z (\Delta x)^2 \widehat{C(\hat{\beta}_x,\hat{\beta}_{xz})}$, that is, we multiply the estimated variance of the estimated marginal effect, $\widehat{V(\Delta \hat{y}/\Delta x)} = \widehat{V(\hat{\beta}_x)} + z^2\widehat{V(\hat{\beta}_{xz})} + 2z\widehat{C(\hat{\beta}_x,\hat{\beta}_{xz})}$, by $(\Delta x)^2$.

We can use the estimates in table 22 to determine the effect of a *Runoff* at various values of *Group* using the difference method simply by subtracting the first from the fourth column, that is, $(\hat{y} \mid Groups,$ $Runoff = 1) - (\hat{y} \mid Groups, Runoff = 0)$. Recall that the case of $(x_c - x_a) = 1$ produces exactly the same results as the derivative method; the differences in predicted values between systems with runoffs and without runoffs, at given values of *Groups* (and the corresponding estimates of uncertainty around those differences in predicted values), appear in table 15.

More generally, for binary variables like our *Runoff*, *Gender*, *Partisanship*, or *Party Discipline*, the only discrete changes meriting consideration are unit increases or decreases, $\Delta x = \pm 1$, and so the estimated marginal effects and confidence intervals plotted in figures 4, 5, 7, 15, and 17 are all identical to the estimated conditional effects of and confidence intervals for a positive switch in the value of that binary indicator. Similarly, the estimated marginal effects of *Groups*, *NP*, and *PS* and confidence intervals plotted in figures 6, 8, 9, 13, 14, and 16 are all identical to the estimated effects and confidence intervals for unit increases in those (nonbinary) variables. If we had wanted to present the estimated effects of, say, a 10 percent rather than a unit (1 percent) increase in parliamentary support, for example, we would simply have multiplied the

---

52. In fact, in our "log-transformed" model of government duration from table 8, the effect of a unit increase in the number of governing parties, which itself enters the model linearly, is linear in the natural log of parliamentary support, and so, had ln(*PS*) been the *x*-axis of figure 12, the same would apply for that presentation.

conditional effect line in figures 9, 14, and 16 by 10, the variance in the formula for the confidence intervals associated with those lines by 10^2, and relabeled the figure as "Effect of a 10% Increase in *Parliamentary Support . . .*"

Graphs of the estimated effects of discrete changes would therefore simply rescale the marginal-effect graphs already shown. We can demonstrate this formally for the standard linear-interaction model as follows. The difference between $\hat{y}_a$ and $\hat{y}_c$, that is, $\hat{y}$ at $x = x_a$ subtracted from $\hat{y}$ at $x = x_c$, is

$$\hat{y}_c - \hat{y}_a = \hat{\gamma}_0 + \hat{\beta}_x x_c + \hat{\beta}_z z_0 + \hat{\beta}_{xz} x_c z_0 - (\hat{\gamma}_0 + \hat{\beta}_x x_a + \hat{\beta}_z z_0 + \hat{\beta}_{xz} x_a z_0)$$

$$= \hat{\beta}_x (x_c - x_a) + \hat{\beta}_{xz} z_0 (x_c - x_a)$$

$$= (x_c - x_a)(\hat{\beta}_x + \hat{\beta}_{xz} z_0)$$

The variance of that difference is then

$$V(\hat{y}_c - \hat{y}_a) = V[(x_c - x_a)(\hat{\beta}_x + \hat{\beta}_{xz} z_0)]$$

$$= (x_c - x_a)^2 V[(\hat{\beta}_x + \hat{\beta}_{xz} z_0)]$$

$$= (x_c - x_a)^2 [V(\hat{\beta}_x) + z_0^2 V(\hat{\beta}_{xz}) + 2 z_0 C(\hat{\beta}_x, \hat{\beta}_{xz})]$$

So, in the case of $(x_c - x_a) = 1$, we have exactly the same results as the derivative method, and in the case of $(x_c - x_a) = \Delta x$ we have the same results rescaled multiplicatively by Δx.

We could also tabulate and/or graph the difference in predicted values as the number of societal groups changes, by one unit (say, from *Groups* = 1 to *Groups* = 2 or, equivalently, from *Groups* = 2 to *Groups* = 3), or by two units (from *Groups* = 1 [the sample minimum] to *Groups* = 3 [just above the sample maximum]), by the presence or absence of a runoff. The differences in predicted values that correspond with a one-unit and a two-unit shift in *Groups*, by *Runoff*, appear in table 25. We could also present a graph containing the difference in predicted values associated with a two-unit shift in *Groups*, but since

TABLE 25. Confidence Intervals for Differences in Predicted *Number of Candidates*

	Runoff = 0			*Runoff* = 1		
	$\hat{y}_c - \hat{y}_a$	$s.e.(\hat{y}_c - \hat{y}_a)$	90% Confidence Interval	$\hat{y}_c - \hat{y}_a$	$s.e.(\hat{y}_c - \hat{y}_a)$	90% Confidence Interval
$\Delta Groups = 1$	−0.979	0.770	[−2.352, 0.394]	1.026	0.540	[0.064, 1.988]
$\Delta Groups = 2$	−1.958	1.541	[−4.704, 0.787]	2.052	1.079	[0.129, 3.976]

there are only two points to be graphed, a table like table 25 is just as informative.

When variables enter the regression models nonlinearly, however, as in the quadratic- and log-transformed-*PS* models of tables 7 and 8, the effect of a discrete change from one value of x to another can be quite different than the effect of a marginal (i.e., infinitesimal) change at that x. That is, except for straight lines, *derivatives* and *slopes* differ from *differences*. In figure 2, for example, the marginal effect of parliamentary support at $PS = 50$ (i.e., the derivative or slope at that point) is $\partial \widehat{GD}/\partial PS = \hat{\beta}_{ps} + 2\hat{\beta}_{ps^2} \, 50 = \hat{\beta}_{ps} + 100 \times \hat{\beta}_{ps^2}$. The effect of a unit change from $PS = 50$ to 51 would be $\Delta \widehat{GD}/\Delta PS = (\hat{\beta}_{ps} \, 51 + \hat{\beta}_{ps^2} \, 51^2) - (\hat{\beta}_{ps} \, 50 + \hat{\beta}_{ps^2} \, 50^2) = \hat{\beta}_{ps} + \hat{\beta}_{ps^2} (51^2 - 50^2) = \hat{\beta}_{ps} + \hat{\beta}_{ps^2} \, 101.$[53] The estimated variances would differ accordingly. The estimated effect of a 10 percent increase from $PS = 45$ percent to 55 percent, $\Delta \widehat{GD}/\Delta PS = (\hat{\beta}_{ps} \, 55 + \hat{\beta}_{ps^2} \, 55^2) - (\hat{\beta}_{ps} \, 45 + \hat{\beta}_{ps^2} \, 45^2) = \hat{\beta}_{ps} \, 10 + \hat{\beta}_{ps^2} \, 1{,}000$, is likewise not equal to the slope at $PS = 45$, $\partial \widehat{GD}/\partial PS = \hat{\beta}_{ps} + 90 \, \hat{\beta}_{ps^2}$, and their standard errors differ also.

Similarly in the log-transformed-*PS* model, the effects of discrete changes in parliamentary support depend not only on the number of governing parties but on the magnitudes and values of those changes in *PS*. The estimated effect of a 10 percent increase, from $PS = 45$ percent to $PS = 55$ percent, and its standard error would be

$$(\widehat{GD}|PS = 55) - (\widehat{GD}|PS = 45) = \hat{\beta}_{\ln(ps)} \, (\ln 55 - \ln 45) + \hat{\beta}_{np\ln(ps)} NP$$
$$\times \, (\ln 55 - \ln 45)$$

$$\overline{V((\widehat{GD}|PS = 55) - (\widehat{GD}|PS = 45))} = (\ln 55 - \ln 45)^2$$
$$\times \, (\widehat{V(\hat{\beta}_{\ln(ps)})} + NP^2 \times \widehat{V(\hat{\beta}_{np\ln(ps)})}$$
$$+ \, 2NP \times \widehat{C(\hat{\beta}_{\ln(ps)}, \hat{\beta}_{np\ln(ps)})})$$

As these two examples illustrate, the conditional effect of a discrete change in an independent variable that enters an interaction model nonlinearly depends not only on the values of the variables with which it interacts but also on the magnitude of the change and from what starting point. We would not, therefore, generally recommend graphing but rather recommend tabulating sets of estimates like these for consideration and discussion.

---

53. We can focus only on the terms that would involve Δx because the rest of the equation drops from these differences.

Distinguishing between Conditional Effects and Predicted Values

How do tables and graphs of conditional effects and those of predicted values differ? Both reveal information about the relation of x to y and how this relationship changes as z varies but in slightly different ways. Graphs and tables of derivatives of or differences in predicted values show directly how the *effect* of x on y changes as z changes. Graphs and tables of predicted values show how the *level* of $\hat{y}$, that is, the prediction for y, changes as x changes, at particular levels of z. By comparing several of these predicted levels, one can also grasp the effects of x or of z on y and how they change as the other variable changes, but, in predicted-level tables and figures, the comparison of effects is less direct and the uncertainties related refer to individual predictions and not to these differences, that is, not to the effects. Selection of one type of table or graph over the other therefore largely depends on the researcher's presentational goals. Either method can effectively convey the substantive results from empirical models involving interactive terms; we stress, however, that either sort of table or graph should incorporate measures of uncertainty into its presentation.

4

THE MEANING, USE, AND ABUSE
OF SOME COMMON
GENERAL-PRACTICE RULES

Having discussed formulation of interactive hypotheses, and interpretation and presentation of effects, we turn now to clarify some general-practice rules often applied in the social sciences.

Colinearity and Mean-Centering the Components
of Interaction Terms

One common concern regarding the estimation of interactive models is the (*multi*)*colinearity*, or high correlation, among independent variables induced by multiplying regressors together to create additional regressors. Colinearity, as social scientists well know, induces large standard errors, reflecting our low confidence in the individual coefficients estimated on these highly correlated factors. What is sometimes forgotten is that these large standard errors are *correctly* large; the effect of x controlling for other terms (i.e., holding them constant) *is* hard to determine with much certainty if x and other terms correlate highly. These large standard errors accurately reflect our high degree of uncertainty in these conditions. These perhaps unfortunate, but very real, facts regarding colinearity led Althauser (1971), for example, to argue against the use of interactive terms at all. However, to omit interactions simply because including them invites a greater degree of uncertainty in parameter estimates is to misspecify intentionally our theoretical propositions. This assures at least inefficiency but most likely induces bias due to standard omitted-variable-bias considerations: namely, if the omitted factor, xz, (partially) correlates with the included factor, x, and (partially) correlates with the dependent variable, y, then bias results. The sign and magnitude of the bias are given by the product of these two partial coefficients.

Scholars therefore struggled valiantly for some technical artifice to reduce interaction-induced colinearity. However, the problem of colinearity is "too little information." As such, the only routes around the problem available to researchers are to ask the data questions that require less information (e.g., only first-order questions, like those in table 10 or table 12) or to obtain more information by drawing new data (preferably less correlated, but more data will help regardless) or by relying more heavily upon the theoretical arguments/assumptions to specify models that ask more precise questions of the data than do generic linear-interactive models (e.g., Franzese 1999, 2002, 2003a).

Scholars have instead devoted inordinate attention to illusory colinearity "cures." The most commonly prescribed "cure" is to "center" the variables (i.e., subtract their sample means or "mean-deviate" them) that comprise the interactive terms. Smith and Sasaki (1979) offered centering as a technique that would improve substantive interpretation of the individual coefficients, and we agree that it might facilitate interpretation in some substantive contexts. Tate (1984) argued that, although centering should not change the substantive effects (actually, it *will* not: see the discussion that follows), it "may improve conditioning through reduction of colinearity" (253). Others, including Morris, Sherman, and Mansfield (1986) and Dunlap and Kemery (1987), recommend centering less circumspectly. The centering technique of Cronbach (1987) has attained considerable acceptance in social science, perhaps due to the promotion of it by Jaccard, Turrisi, and Wan (1990). Unfortunately, Cronbach's clarification on the extremely limited value of centering seems less widely known.

To be sure, the centering procedure of Cronbach (1987) is harmless; however, it also offers no help against the "too little information" problem of colinearity, if understood correctly. Our concern is that centering seems widely misunderstood and misinterpreted. Some existing scholarly research claims, wrongly, that centering helps evade colinearity in some manner that actually produces more certain effect estimates. Centering adds no new information of any sort to the empirical estimation, and so it cannot possibly produce more precise estimates. Centering merely changes the substantive question to which the coefficients and t-tests of those coefficients refer.

Consider this standard linear-interactive model:

$$y = \beta_0 + \beta_x x + \beta_z z + \beta_{xz} xz + \varepsilon \qquad (31)$$

Cronbach (1987) suggested subtracting the sample means from each of the independent variables involved in the interaction and multiplying the resulting demeaned variables together for the interaction term. The mean-centered model, then (using γ to represent coefficient values resulting from use of the centered data), is as follows:

$$y = \gamma_{0*} + \gamma_{x*} x^* + \gamma_{z*} z^* + \gamma_{x*z*} x^* z^* + \varepsilon^* \qquad (32)$$

where $x^* = x - \bar{x}$ and $z^* = z - \bar{z}$.

Cronbach (1987) argued that rescaling the variables thusly could insure against computational errors—that is, errors that are literally computational: deriving from inescapable rounding errors in translating from computer binary to human base-ten—that severe colinearity might induce.[1] Cronbach (1987) also noted that centered and noncentered models "are logically interchangeable, and under most circumstances it makes no difference which is used" (415). Given the many thousands of times computing precision has increased since Cronbach's writing, the computational concern has no current practical relevance in social science, and so it now makes no difference under any circumstances.

Because centering does not affect the substance of any empirical estimation in any way, because it will not affect the computational algorithms of any modern statistical software, and because it is so widely misunderstood in the field, we join Friedrich (1982), Southwood (1978),

---

1. The computational issue here involves matrix inversion, namely, the $(X'X)^{-1}$ in OLS formulas for coefficient and standard-error estimates, some of whose columns (i.e., independent variables) may correlate nearly perfectly. If columns of X correlate perfectly, the determinant of $(X'X)$, which appears in the denominator of the formula for $(X'X)^{-1}$, is zero. Division by zero is, of course, impossible; therefore, obtaining distinct coefficient estimates (and thus standard errors) when (some) columns of X correlate perfectly is impossible. All modern regression software warns of perfect colinearity when it obtains a zero determinant before allowing the computer to crash trying to divide by zero. Most warn of near-perfect colinearity well short of obtaining identically zero for that critical determinant, that is, well short of perfect colinearity, because the translation from the base-ten data matrix to the binary of computers involves rounding error. When something near zero appears in a denominator and contains slight rounding error, the final answer could exhibit massive error. This is the concern that Cronbach raised. The multiplicative terms in interactive regressions, he feared, could be near enough to perfect colinearity to cause severe binary-to-base-ten rounding-error problems. However, since his writing, computers have become many thousands of times more exact in their binary calculations' approximation to base ten, meaning that even this computational concern is no longer present in any practical social-science context.

and others in strongly advising the abandonment of the practice or, at least, far greater care in interpreting and presenting the results following its implementation. To clarify what centering does to the numeric and substantive estimates of an interactive analysis, which is something and nothing, respectively, consider again our basic linear-interaction model and its centered version, which appear in equations (31) and (32), respectively. Starting from equation (32), and substituting terms, we see that

$$y = \gamma_{0^*} + \gamma_{x^*}(x - \bar{x}) + \gamma_{z^*}(z - \bar{z}) + \gamma_{x^*z^*}(x - \bar{x})(z - \bar{z}) + \varepsilon^* \qquad (33)$$

$$y = \gamma_{0^*} + \gamma_{x^*}x - \gamma_{x^*}\bar{x} + \gamma_{z^*}z - \gamma_{z^*}\bar{z} + \gamma_{x^*z^*}xz - \gamma_{x^*z^*}\bar{x}z - \gamma_{x^*z^*}x\bar{z}$$
$$+ \gamma_{x^*z^*}\overline{xz} + \varepsilon^*$$

$$y = (\gamma_{0^*} - \gamma_{x^*}\bar{x} - \gamma_{z^*}\bar{z} + \gamma_{x^*z^*}\overline{xz}) + (\gamma_{x^*} - \gamma_{x^*z^*}\bar{z})x + (\gamma_{z^*} - \gamma_{x^*z^*}\bar{x})z$$
$$+ \gamma_{x^*z^*}xz + \varepsilon^* \qquad (34)$$

Comparing the centered equation in (34) with the original model in (31) highlights the exact correspondence of results between the centered and uncentered regression models:

$$\beta_0 = \gamma_{0^*} - \gamma_{x^*}\bar{x} - \gamma_{z^*}\bar{z} + \gamma_{x^*z^*}\overline{xz}$$

$$\beta_x = \gamma_{x^*} - \gamma_{x^*z^*}\bar{z}$$

$$\beta_z = \gamma_{z^*} - \gamma_{x^*z^*}\bar{x}$$

$$\beta_{xz} = \gamma_{x^*z^*}$$

Collecting terms, we see that the first parenthetical expression in equation (34) contains its set of constant terms and thus equals the intercept, β_0, from (31). The second parenthetical expression in (34) is its ultimate coefficient on x, which is equal to β_x from (31), and the third parenthetical expression is the ultimate coefficient on z in (34), which equals β_z in (31). The fourth term is the coefficient on xz in each model. Trivially, since the right-hand-side models are mathematically interchangeable, the estimated residuals and therefore the estimated residual variance from the centered and uncentered models are also identical.

As we explained previously, researchers' common troubles arise when they confuse coefficients with effects. We know, for example, that the marginal effect of x on y in equation (31) would be $\partial y/\partial x = \beta_x + \beta_{xz}z$. The marginal effect of x^* on y given equation (32) would be $\partial y/\partial x^* = \gamma_{x^*} + \gamma_{x^*z^*}z^*$. Since $\beta_x = \gamma_{x^*} - \gamma_{x^*z^*}\bar{z}$, we can express $\gamma_{x^*} = \beta_x + \gamma_{x^*z^*}\bar{z}$. Therefore

$$\partial y/\partial x^* = \beta_x + \gamma_{x^*z^*}\bar{z} + \gamma_{x^*z^*}z^*$$

Then, given that $z^* = z - \bar{z}$, we have

$$\partial y/\partial x^* = \beta_x + \gamma_{x^*z^*}\bar{z} + \gamma_{x^*z^*}z - \gamma_{x^*z^*}\bar{z} = \beta_x + \gamma_{x^*z^*}z$$

Finally, since $\beta_{xz} = \gamma_{x^*z^*}$, we conclude

$$\partial y/\partial x^* = \beta_x + \beta_{xz}z = \partial y/\partial x$$

Stated directly, the point is obvious: the effect of a marginal increase in the centered version of x is identical to the effect of a marginal increase in uncentered x. The same identity applies to the effects of z, of course. We reiterate: centering does not change the estimated *effects* of any variables.

Further, the estimated variance-covariances (i.e., standard errors, etc.) of those effects are also identical. Thus, the estimated statistical certainty of the estimated effects is also unchanged by centering. For the uncentered data, $V(\partial\hat{y}/\partial x) = V(\hat{\beta}_x) + z^2 V(\hat{\beta}_{xz}) + 2zC(\hat{\beta}_x,\hat{\beta}_{xz})$. Using the mean-centered model:

$$V(\partial\hat{y}/\partial x^*) = V(\hat{\gamma}_{x^*}) + (z^*)^2 V(\hat{\gamma}_{x^*z^*}) + 2z^* C(\hat{\gamma}_{x^*},\hat{\gamma}_{x^*z^*})$$

Substituting $\hat{\gamma}_{x^*} = \hat{\beta}_x + \hat{\gamma}_{x^*z^*}\bar{z}$ and $\hat{\gamma}_{x^*z^*} = \hat{\beta}_{xz}$

$$V(\partial\hat{y}/\partial x^*) = V(\hat{\beta}_x + \hat{\beta}_{xz}\bar{z}) + (z^*)^2 V(\hat{\beta}_{xz}) + 2z^* C(\hat{\beta}_x + \hat{\beta}_{xz}\bar{z}, \hat{\beta}_{xz})$$

$$V(\partial\hat{y}/\partial x^*) = V(\hat{\beta}_x) + \bar{z}^2 V(\hat{\beta}_{xz}) + 2\bar{z}C(\hat{\beta}_x,\hat{\beta}_{xz}) + (z^*)^2 V(\hat{\beta}_{xz})$$
$$+ 2z^* C(\hat{\beta}_x + \hat{\beta}_{xz}\bar{z},\hat{\beta}_{xz})$$

Rearranging terms and substituting $z^* = z - \bar{z}$:

$$V(\partial\hat{y}/\partial x^*) = V(\hat{\beta}_x) + \bar{z}^2 V(\hat{\beta}_{xz}) + 2\bar{z}C(\hat{\beta}_x,\hat{\beta}_{xz})$$
$$+ (z - \bar{z})^2 V(\hat{\beta}_{xz}) + 2(z - \bar{z})C(\hat{\beta}_x,\hat{\beta}_{xz}) + 2(z - \bar{z})\bar{z} V(\hat{\beta}_{xz})$$

$$V(\partial\hat{y}/\partial x^*) = V(\hat{\beta}_x) + z^2 V(\hat{\beta}_{xz}) + 2zC(\hat{\beta}_x,\hat{\beta}_{xz}) = V(\partial\hat{y}/\partial x)$$

The variances of the estimated marginal effects of the centered x and of the uncentered x are identical. The same holds for the variances of the estimated marginal effects of z and mean-centered z, of course. As with the coefficients, the numeric values of the elements in the variance-covariance matrices for the coefficients using uncentered and centered data will naturally differ from each other, but exact correspondence in the estimated effects and the estimated variances of effects can be derived through algebraic manipulation of these values. As an example, recall that $\beta_x = \gamma_{x^*} - \gamma_{x^*z^*}\bar{z}$. This implies that $V(\hat{\beta}_x) = V(\hat{\gamma}_{x^*} - \hat{\gamma}_{x^*z^*}\bar{z}) = V(\hat{\gamma}_{x^*})$

$+ \bar{z}^2 V(\hat{\gamma}_{x^*z^*}) - 2\bar{z}C(\hat{\gamma}_{x^*}, \hat{\gamma}_{x^*z^*})$. Hence, while the estimated *coefficients* and variance-covariance matrices of *coefficients* will differ numerically (i.e., $\hat{\beta}_x \neq \hat{\gamma}_{x^*}$ and $\widehat{V(\hat{\beta}_x)} \neq \widehat{V(\hat{\gamma}_{x^*})}$), the estimated *effects* and the precision of the estimated *effects* of the variables will be identical, regardless of whether the data are centered or uncentered. Again, we warn the reader against confusing *coefficients* with *effects*.

If all estimates of the substantive effects and all estimates of the certainty of those substantive effects are identical whether the data are mean-deviated or left uncentered, how, one might wonder, can some key coefficient estimates, their standard errors, and the corresponding *t*-statistics differ? The answer is simply that the coefficients and associated standard errors and *t*-statistics do not refer to the effects at the same substantive values of the regressors across centered and uncentered models. For example, in our standard model, $y = \beta_0 + \beta_x x + \beta_z z + \beta_{xz} xz + \varepsilon$, the coefficient β_x gives the effect of a unit increase in x when z equals zero; its standard error and the resulting *t*-ratio refer to the certainty of that x effect at that particular z value. In $y = \gamma_{0^*} + \gamma_{x^*} x^* + \gamma_{z^*} z^* + \gamma_{x^*z^*} x^* z^* + \varepsilon^*$, the coefficient γ_{x^*} gives the effect of a unit increase in x^* (or x, since a unit increase in x or x^* is the same thing) when z^* equals zero, which is not at all the same value as when $z = 0$ (assuming, of course, that $\bar{z} \neq 0$). Since $z^* = z - \bar{z}$, the centered z^* equals zero when the uncentered z equals its mean, not when $z = 0$ (except in the specific case where $\bar{z} = 0$). The standard error of this coefficient estimate, $\hat{\gamma}_{x^*}$, and the resulting *t*-ratio also refer to the certainty of the effect of a one-unit change in x at this *different* $z = \bar{z}$ value. Coefficients, standard errors, and *t*-statistics differ in the centered and the noncentered models because they refer to different substantive quantities, not because either model produces different, much less any better, estimates of effects than does the other.

Centering can, in this manner, actually be useful for substantive interpretation in some contexts. If interpreted carefully and understood fully, centering sometimes can facilitate a more substantively grounded discussion of the empirical analysis. If z cannot logically equal zero, then substantive interpretation of β_x is vacuous, but examining the effect of x when z is equal to its sample mean might be substantively revealing. If so, researchers might advantageously center z around its mean to aid substantive interpretation and discussion of β_x. That is, centering z around its mean allows one to interpret the coefficient on x as the effect of x when z equals its mean rather than when z equals zero. Further, it allows the researcher to interpret the *t*-statistic on $\hat{\gamma}_x$ as the statistical significance of x when z happens to equal its mean, which may likewise simplify discussion in some contexts.

Accordingly, our concern is that researchers too often misinterpret the results of centering—and have come to the mistaken conclusion that centering alters the estimates of effects or the estimated significance of effects. We recommend that centering transformations, if applied at all, be applied only with the aim to improve substantive presentation, not, mistakenly, to improve (apparent) statistical precision and certainly not, reprehensibly, to move the value of z to which the standard t-ratio refers so as to maximize the number of asterisks of statistical significance on reported t-tests. The substantive interpretation of the effects and the certainty of those effects are completely unaffected by this algebraic sleight-of-hand.

Including x and z when xz Appears

To estimate models containing multiplicative interaction terms, most texts advise a hierarchical testing procedure: that is, if xz enters the model, then x and z must also. If wxz appears, then all (six) of the lower order combinations (x, w, z, xw, xz, wz) must appear also, and so on analogously for higher order interactions. Allison (1979), for example, writes, "[The] common rule . . . is that testing for interaction in multiple regression should only be done hierarchically . . . If a rationale for this rule is given at all, it is usually that additive relationships somehow have priority over multiplicative relationships" (149–50). This rule is probably an advisable one, if researchers must have a rule. Certainly it is a much safer rule than an alternative proviso that one can include or not include components to interactions with little concern or consideration. However, we believe that researchers must understand the logical foundations of the models they estimate and the meaning and purpose of any proffered rule, instead of merely following such rules by rote. We argue instead for theoretically driven empirical specifications with better appreciation of the assumptions underlying alternative models. While the rule of including x and z if including xz may be a quite reasonable application of Occam's razor and is often practically advisable, it is neither logically nor statistically strictly necessary.

As proof that the rule is not logically necessary, notice that one can decompose any variable into the product of two or more others; therefore, strict adherence to this rule would actually entail infinite regress. As a substantive example, note that real GDP (per capita) equals nominal GDP times a price-index deflator (times the population inverse); conversely, nominal GDP (per capita) is real GDP times a price index (times the population inverse). Nothing statistically or logically requires researchers to include all of these components in every model containing some subset of them. Researchers should, instead, estimate the models their theories suggest.

That said, several good reasons to follow the rule exist. First, given the state of social-science theory, the models implied by theory will often be insufficiently specified as to whether to include x and/or z in an interactive model. Due scientific caution would then suggest including x and z to allow the simpler linear-additive theory a chance. (This is Occam's razor.) Failing to do so would tend to yield falsely significant estimates of coefficients on xz if, in fact, x or z or both had just linear-additive effect on y. Second, inclusion of the x and z terms in models involving xz allows a nonzero intercept to the conditional effect lines, such as those plotted in chapter 3. This is important because, even if the effect of x on y is truly zero when z is zero, if this conditional relationship is nonlinear, allowing a nonzero intercept to the linear-interactive estimates of the truly nonlinear interaction (by including x and z) will enhance the accuracy of the linear approximation. Third, and perhaps most important, even when the theory clearly excludes x and/or z from the model, that is, when it unequivocally establishes the effect of one (or both) variable(s) to be zero when the other is zero, the researcher can and should test that prediction and report the certainty with which the data support the exclusion. If that test supports exclusion, then both theory and evidence recommend exclusion of the components, and continued inclusion would be the misspecification of the model. For this sort of empirical exploration, only finding a coefficient expected to be zero in fact to be estimated as (very close to) zero and, highly preferably, with small standard error is clear evidence from the data that the assumption holds. That is, clearest support for the assumption comes from failure to reject *because the estimate is with considerable certainty near zero* rather than because the estimate has very large standard error. In sum, then, this rule, as an application of Occam's razor, is a safer adage than its opposite, but researchers should still, first, understand the basis for the rule and, second, should not shy from breaking it if their theory *and the data* strongly suggest doing so.

We now elaborate these points more fully and formally. If the theory expressly excludes z from having any effect on y when x is zero—that is, nonzero presence of x is a necessary condition for z to affect y, the correct model is

$$y = \beta_0 + \beta_x x + \beta_{xz} xz + \varepsilon \tag{35}$$

By this model, as theory demands, the effect of z on y, $\partial y/\partial z$, equals $\beta_{xz} x$, which is zero when $x = 0$. Estimating this model assumes that x

must be present for z to affect y but does not allow the data to adjudicate the question. If z does affect y even when x is zero, equation (35) would suffer omitted-variable bias, with coefficient estimates wrongly attributing the omitted variable's effects to the variable(s) that do enter the model and that correlate with the omissions. In this case, the omission will most likely imply a biased β_{xz} estimate (primarily). Regression estimation will attribute some of the true-but-omitted effect of z when $x = 0$ to z's interaction with x, and so the estimate of β_{xz} will be too large (small) when this true-but-omitted effect is positive (negative). Thus, if the omitted effect is positive, the estimated effect of z on y ($\partial\hat{y}/\partial z = \hat{\beta}_{xz}x$) will reflect a greater conditional effect than truly exists (i.e., greater slope to this effect line), with underestimation of the effect of z on y at low values of x and overestimation at high values of x. Conversely, the effect of x on y ($\partial\hat{y}/\partial x = \hat{\beta}_x + \hat{\beta}_{xz}z$) will be estimated as more conditional upon z than it truly is, implying too great a slope to this effect line and, likely, also too low an intercept ($\hat{\beta}_x$) to that effect line.

Rather than assume such necessity clauses by omitting key interaction components, we suggest that researchers test them by first estimating the model including all lower order components:

$$y = \beta_0 + \beta_x x + \beta_z z + \beta_{xz} xz + \varepsilon \qquad (36)$$

An insignificant coefficient of β_z here might then support the exclusion theory and provide some justification for proceeding with the necessity clause in place. But recall that a t-test on $\hat{\beta}_z$ only refers to the effect of z when x equals zero. The theory concludes that β_z should equal zero, and so we would hardly want to accept that hypothesis merely because we fail to reject it at some generous significance level like $p < 0.10$. Recall that failure to reject can occur with small coefficient estimates and small standard errors, small coefficient estimates and large standard errors, or large coefficient estimates and larger standard errors. Only the first of these should give the researcher great comfort that he or she may estimate the model that assumes the necessity clause by omitting (an) interaction component(s); the second gives less support for such a restriction; and the last gives very little or none at all.

In summary, estimating models like (36) that include all interaction components when true models, for instance, (35), actually exclude them will cost researchers some inefficiency if not bias. Estimating (36) when the true model is (35) involves trying to estimate more coefficients than necessary, which implies inflated standard errors. Moreover, these included-but-unnecessary coefficients, β_x or β_z, are on variables, x or z, that are likely highly correlated with the necessary ones, xz, which implies greatly inflated standard errors. Thus, the inefficiency of overcautious interaction-component inclusion could easily and often be severe enough to lead researchers to miss many interactions actually present in their subject. Especially as theory advances to grapple with the complex conditionality of the subjects that social scientists study, and as empirical models attempt to follow even though data remain stubbornly scarce, such inefficiency can very easily become unaffordable. Thus, we recommend that researchers (a) acknowledge and discuss the assumptions/arguments underlying the decision to omit *or to include* components of their interaction terms, (b) gauge statistically the certainty with which the data support those assumptions, and then (c) apply Occam's razor by following hierarchical procedures unless theory and data clearly indicate that doing so is unnecessary and overly cautious.

5

EXTENSIONS

We turn next to some more technical statistical concerns often raised regarding interaction-term usage in regression analysis. The first issue regards separate-sample versus pooled-sample estimation of interactive effects. The second issue concerns estimation and interpretation of interaction terms in nonlinear models, including qualitative dependent-variable models like logit and probit models of binary outcomes. The third issue concerns modeling and estimating stochastically (rather than determinately) interactive relationships.[1]

Separate-Sample versus Pooled-Sample Estimation of Interactive Effects

Researchers often explore the interactive effects of nominal (binary, categorical, etc.) variables by splitting their samples according to these categories and estimating the same model separately in each subsample.[2] In behavioral research, for example, scholars may analyze interactive hypotheses that individual characteristics structure the impact of other variables by estimating the same model in subsamples separated by race,

---

1. See Franzese (2005) for further, formal discussion of the first and third issues.

2. Indeed, sometimes even ordinal or cardinal variables are separated into high(er) and low(er) categories for sample splitting in this manner. In addition to the considerations to be discussed in this section, this will typically entail inefficiency as the gradations of ordinal or cardinal information are discarded in the conversion to nominal categorization, although the practice may be justifiable in some cases on other grounds.

gender, and so on. A researcher might, for instance, estimate the effect of socioeconomic status on political participation separately in samples of male and female respondents to explore whether socioeconomic status affects the propensity to vote differently by gender. In comparative or international politics too, researchers might estimate the same model separately by country or region to explore whether national or regional contexts condition the effects of key variables. A political economist might, for instance, estimate a model of electoral cycles in monetary policy separately in subsamples of fixed- and flexible-exchange-rate country times. Similar subsample estimation strategies populate all subfields of political science and other social sciences.

Such subsample estimation (1) produces valid estimates of the (conditional) effects of the other variables at these different values of the "moderating" variable, (2) commendably recognizes the conditionality of the underlying arguments, and (3) can (perhaps with some effort) reproduce any of the efficiency and other desirable statistical properties of the alternative strategy of pooling with (nominal) interactions. However, these subsample procedures also isolate, at least presentationally, one variable as the moderator in what is logically a symmetric process—if x moderates the effect of z on y, then z moderates the effect of x on y and vice versa—thereby obscuring the converse. More fundamentally, these procedures do not facilitate statistical comparison of the effects of "moderated" or "moderating" variables; that is, one cannot as easily determine whether any differences in estimated effects across subsamples are statistically significant or as easily determine the (conditional) effects of the variable being treated as the moderating variable as one can in the pooling strategy.

An alternative approach is to estimate a model that keeps the subsamples together and that includes interaction terms of all of the other covariates, including the constant, with the variable being treated as the moderator; this is sometimes called a "fully dummy-interactive" model. The two approaches (separate sample versus fully dummy interactive pooled sample) extract almost identical sets of information from the data, but pooled-sample estimation extracts slightly more, potentially more efficiently, and more easily allows statistical testing of the full set of typical interactive hypotheses. That is, any desirable statistical properties that one can achieve by one strategy can, perhaps with considerable effort, be achieved by the other (see, e.g., Jusko and Shively 2005). However, we believe that the pooled interactive strategy lends itself more easily to obtaining these desirable qualities and, in some cases, also to presenting and interpreting results. Hence, we suggest that separate-sample estimation be reserved for exploratory and sensitivity and robustness consideration stages of analysis. We recommend pooled-sample approaches for final analysis and presentation. In either case, however, we note that theory should dictate the use of fully interactive (or separate-subsample) versus selectively interactive models. We do not advocate that fully interactive models or separate-sample models be used as a substitute for theoretically informed specifications. However, if a researcher is intent on "splitting the samples," then estimation using a fully interactive pooled model is a better alternative to separate-sample estimation.

As an example, a researcher, wishing to explore gender differences, g, in the effect of socioeconomic status and other independent variables, $\mathbf{X}$, on propensity to vote, y, separates the sample into males and females and estimates

$$\textit{Sample } g = \textit{Male: } \mathbf{y_m} = \mathbf{X\beta_m} + \mathbf{u_m} \tag{37}$$

$$\textit{Sample } g = \textit{Female: } \mathbf{y_f} = \mathbf{X\beta_f} + \mathbf{u_f} \tag{38}$$

Let M (F) be the number of observations in the male (female) sample. Let k index the columns of $\mathbf{X}$ (e.g., $\mathbf{x}_{gk}$ represents the kth independent variable for the gender g sample; β_{gk} is the coefficient on that kth independent variable for that gender g sample) and let K be the number of independent variables (excluding the constant). To obtain distinct coefficient estimates by gender, the researcher has several options.

Most easily, the researcher could estimate models (37) and (38) separately, once per subsample. Or, he or she could pool the data into one sample and reconfigure the $\mathbf{X}$ matrix by manually creating separate $\mathbf{X_m}$ and $\mathbf{X_f}$ variables for each column of $\mathbf{X}$, where $\mathbf{X_m}$ replaces each female respondent's $\mathbf{X}$ value with zero and $\mathbf{X_f}$ does so for male respondents. This allows distinct coefficients on $\mathbf{X_m}$ and $\mathbf{X_f}$ and, if the constant (intercept) is also separated into $\mathbf{X_m}$ and $\mathbf{X_f}$ in this way, will produce exactly the same coefficient estimates as separate-sample estimation does. Identically to this manual procedure, the researcher could simply create an indicator variable for $g_m = \textit{Male}$ and another indicator for $g_f = \textit{Female}$ and include these two indicators in place of the intercept and the interaction of each of these indicators with all of the other independent variables in place of those independent variables. Each $g_m\mathbf{X}$ and $g_f\mathbf{X}$ here will equal the $\mathbf{X_m}$ and $\mathbf{X_f}$ from the manual procedure just described, and so this also produces exactly the same coefficient estimates as the separate-sample estimation. Finally, the researcher could simply create one gender indicator, say, the female g_f, and include in the pooled-sample estimation all of the $\mathbf{X}$ independent variables (including the constant), unmodified, plus that g_f

indicator times each of these **X** variables (including the constant, which product just reproduces g_f). This, too, would produce the same substantive estimates for the model as separate-sample estimation, but the coefficients would now refer to different aspects of that substance. The coefficient on each variable x_k (including the intercept) in this last option would refer to the effect on y of that variable among males, whereas those coefficients on each x_k *plus the coefficient on the corresponding interaction term,* $g_f x_k$, would refer to the effect on y of that x_k among females. And the coefficient on $g_f x_k$ would refer to the difference in the effect of that x_k among females and the effect of that x_k among males. If all of these approaches produce the same substantive results from their estimates, why might researchers prefer one or the other of them?

In our review, researchers rarely offer reasons for presenting separate subsample estimations of interactive effects. Perhaps some do not realize that pooled-sample alternatives using interaction terms exist and, as we show next, are at least equivalent on all grounds except, perhaps, convenience. Others may note more explicitly that, lacking a priori hypotheses about what differences in the effects of the various x_k to expect across their subsamples, they wish simply to explore inductively what some possible candidates for interactive effects might be, and they find separate-sample estimation a convenient and easily interpreted means of conducting such exploration. The more technically savvy might even suggest that they did not wish to impose or estimate any distributional features of the residual term across subsamples, which would be necessary to validate statistical comparison of subsample coefficient estimates in pooled estimation.

In the separate-sample approach, researchers estimate one equation for males:

$$\begin{bmatrix} y_{1m} \\ \vdots \\ y_{Mm} \end{bmatrix} = \begin{bmatrix} 1 & X_{m11} & \cdots & X_{mK1} \\ \vdots & \vdots & & \vdots \\ 1 & X_{m1M} & \cdots & X_{mKM} \end{bmatrix} \begin{bmatrix} \beta_{m0} \\ \beta_{m1} \\ \vdots \\ \beta_{mK} \end{bmatrix} + \begin{bmatrix} \varepsilon_{1m} \\ \vdots \\ \varepsilon_{Mm} \end{bmatrix} \tag{39}$$

and the exactly analogous equation for females. Table 26 provides OLS regression results from conducting this split-sample analysis (using our very simple *Support for Social Welfare* example). Typically, researchers will estimate these equations separately in each subsample and "eyeball" the results for differences in estimated β, which, assuming no other interactions, reflect directly the effect of the associated x in that subsample. This provides the often-cited ease of interpretation in separate-sample estimation. However, the second or third of the pooled-sample options described earlier (i.e., creating distinct X_f and X_m variables manually or by dummy-variable interaction) exactly replicates these separate-subsample coefficient estimates. If researchers prefer this sort of interpretability, pooled-sample estimation can also produce it. Presentationally, too, one can just as easily display two columns of coefficient estimates from one pooled-sample equation as from two separate-sample estimations. Therefore, direct interpretability of effects by subsample cannot adjudicate between pooled-sample and separate-sample approaches since one can present the same results in the same fashion regardless of whether those results derived from pooled-sample or separate-sample estimation.

Underlying any separate-sample estimation in the first place is at least the hunch that the effects of some independent variables differ across the categories distinguished by the subsamples. Thus, certainly, anyone conducting such analysis will wish to compare coefficient estimates across such subsamples. In table 26, a researcher might eyeball the differences in the estimated coefficient for *Republican* in the sample for males, $\hat{\beta}_R = -0.2205$, and in the sample for females, $\hat{\beta}_R = -0.1368$, and conclude (often by some unspoken or, worse, arbitrary standard) that these coefficients look "different enough." *If* classical OLS assumptions apply in

TABLE 26. OLS Regression Results, *Support for Social Welfare*, Pooled and Split Samples

	Pooled Sample Coefficient (standard error) p-Value	Males Only Coefficient (standard error) p-Value	Females Only Coefficient (standard error) p-Value
Female	−0.0031 (0.0144) *0.828*	—	—
Republican	−0.2205 (0.0155) *0.000*	−0.2205 (0.0154) *0.000*	−0.1368 (0.0148) *0.000*
Female × Republican	0.0837 (0.0214) *0.000*	—	—
Intercept	0.7451 (0.0110) *0.000*	0.7451 (0.0109) *0.000*	0.7420 (0.0094) *0.000*
N (*df*)	1,077 (1,073)	498 (496)	579 (577)
Adjusted R^2	0.223	0.290	0.128
P > F	0.000	0.000	0.000

Note: Cell entries are the estimated coefficient, with standard error in parentheses, and two-sided p-level (probability $|T| > t$) referring to the null hypothesis that $\beta = 0$ in italics.

each subsample (the OLS $\hat{\beta}$ are the best linear unbiased estimates [BLUE]), then the researcher could test the statistical significance of any differences in parameters estimated separately across subsamples by difference tests of each $\hat{\beta}_f$ and corresponding $\hat{\beta}_m$:[3]

$$H_0: \beta_f = \beta_m \quad \text{or} \quad \beta_f - \beta_m = 0$$

Conducting the standard t-test of this null hypothesis:

$$\frac{(\hat{\beta}_f - \hat{\beta}_m) - 0}{s.e.\ (\hat{\beta}_f - \hat{\beta}_m)} = \frac{(\hat{\beta}_f - \hat{\beta}_m)}{\sqrt{\widehat{V(\hat{\beta}_f)} + \widehat{V(\hat{\beta}_m)} - 2\widehat{C(\hat{\beta}_f,\hat{\beta}_m)}}} = \frac{(\hat{\beta}_f - \hat{\beta}_m)}{\sqrt{\widehat{V(\hat{\beta}_f)} + \widehat{V(\hat{\beta}_m)}}} \quad (40)$$

The equality of the last expression to the previous two follows in this case, as it would not generally, because $\hat{\beta}_f$ and $\hat{\beta}_m$ will not covary due to the orthogonality of the gender indicators. Using our example, we would thus calculate $((\hat{\beta}_f - \hat{\beta}_m) - 0)/s.e.(\hat{\beta}_f - \hat{\beta}_m) = (-0.1368 - (-0.2205))/$ $\sqrt{(0.0148)^2 + (0.0154)^2} = 0.0837/0.0214 \approx 3.92$. The resulting t-test on this value suggests $p < 0.0001$: these estimated coefficients do appear to be statistically distinguishable from each other.

Few researchers in our review of the literature actually conducted this test; at best, they offered some reference to the *individual* standard errors of the two coefficient estimates in question. The subsample coefficient estimates will be independent by construction (the orthogonality of the indicator variables assures this), but the simple sum of the standard errors of the two coefficients is not the correct standard error for the estimated difference. The standard error of the estimated difference between the two coefficients is the square root of the sum of the estimated variances of the two coefficients. To conduct this comparison across subsamples of estimated effects, the reader should square the reported standard-error estimates, sum those variances, and square-root that sum.

Pooled-sample estimation allows a more directly interpretable formulation if the goal is to test whether effects differ across subsamples. Namely, with the right-hand side of the model specified as $\mathbf{X}$ and the nominal indicator(s) times $\mathbf{X}$, the coefficient(s) on the interaction terms directly reveal the difference in effects across subsamples and the standard t-tests of those interaction-term coefficients directly reveal the statistical significance of those differences in effects.[4] A researcher seeking to deter-

---

3. Researchers may also conduct the joint-hypothesis test that all of the coefficients are equal across subsamples, $H_0: \hat{\mathbf{\beta}}_m = \hat{\mathbf{\beta}}_f$, with a standard F-test: $(\hat{\mathbf{\beta}}_m - \hat{\mathbf{\beta}}_f)'[\widehat{V(\hat{\mathbf{\beta}}_m - \hat{\mathbf{\beta}}_f)}]^{-1}$ $(\hat{\mathbf{\beta}}_m - \hat{\mathbf{\beta}}_f) \sim F_{k,n-k}$.

4. Likewise, the standard F-test on the set of interaction terms tests whether the set of effects of $\mathbf{X}$ jointly differ across subsamples; see note 3.

mine whether the effect of *Republican* differs across females and males would need to calculate $\hat{\beta}_{R,females} - \hat{\beta}_{R,males} = 0.0837$ by subtracting the respective estimated coefficients acquired through separate-sample estimation. The pooled-sample estimation already provides this information, in the estimated coefficient, $\hat{\beta}_{FR} = 0.0837$. Further, instead of calculating the estimated standard error $s.e.(\hat{\beta}_f - \hat{\beta}_m)$ based on the two separate samples, per equation (40), the researcher can determine whether the difference in the effect of *Republican* between females and males is statistically distinguishable from zero by simply conducting a t-test using the results from the pooled-sample estimation: divide the estimated coefficient $\hat{\beta}_{FR}$ by its estimated standard error: $0.0837/0.0214 \approx 3.92$.

Thus, pooled-sample estimation offers two ways of presenting the same substantive results. One way replicates the same interpretability of coefficients as effects in subsamples afforded by separate-sample estimation. Another affords direct interpretation of coefficients as the estimated difference between effects across subsamples, as well as the standard t-tests or F-tests on those coefficients as revealing the statistical significance of that estimated difference. Pooled-sample estimation streamlines the process of testing the substantive hypotheses that researchers often seek to examine.

Moreover, pooling not only produces identical effect estimates as those obtained from separate samples, but it also (under classical linear regression model [CLRM] assumptions) constrains the variance of residuals, s^2, to be equal for the two samples and not to covary across subsamples. Separate-sample estimation makes no such assumptions; thus, pooled-sample estimation borrows strength from the other subsample(s) to obtain better (i.e., more efficient) standard error estimates, although only correctly so if these assumptions are true. Formally, these features are seen most directly for the case where $\mathbf{X}$ is arranged in block diagonal, either manually or by dummy-variable interactions:

$$
\underset{(M+F)\times 1}{\begin{bmatrix} y_{m1} \\ \vdots \\ y_{mM} \\ y_{f1} \\ \vdots \\ y_{mF} \end{bmatrix}} = \underset{(M+F)\times(2K+2)}{\begin{bmatrix} 1 & X_{m11} & \cdots & X_{mK1} & 0 & \cdots & & 0 \\ \vdots & & & \vdots & \vdots & & \ddots & \\ 1 & X_{m1M} & & X_{mKM} & 0 & \cdots & & 0 \\ 0 & \cdots & & 0 & 1 & X_{f11} & \cdots & X_{fK1} \\ & 0 & & & & \vdots & & \\ \vdots & & \ddots & & \vdots & & & \vdots \\ 0 & \cdots & & 0 & 1 & X_{f1F} & & X_{fKF} \end{bmatrix}} \underset{2K+2}{\begin{bmatrix} \beta_{m0} \\ \beta_{m1} \\ \vdots \\ \beta_{mK} \\ \beta_{f0} \\ \beta_{f1} \\ \vdots \\ \beta_{fK} \end{bmatrix}} + \begin{bmatrix} \varepsilon_{m1} \\ \vdots \\ \varepsilon_{mM} \\ \varepsilon_{f1} \\ \varepsilon_{fF} \end{bmatrix} \quad (41)
$$

Recall that $\hat{\Sigma} = s^2 (X'X)^{-1}$. Since the X matrix here is block diagonal, the inverse will also be block diagonal, and the elements for males of $(X'X)^{-1}$ and $X'y$, which comprise the coefficient estimate for males, $\hat{\beta}_m = (X_m'X_m)^{-1} X_m'y_m$, are identical to what they would have been with the samples separated. The statistical test for the equality of the male and female coefficient estimates is then just the standard F-test on the equality of sets of two parameters ($\beta_f = \beta_m$). Note, though, that the single s^2 estimated here naturally differs from the two, s_m^2 and s_f^2, estimated separately in the subsample estimates. Pooled OLS assumes that s^2 is the same across the two samples. That one s^2 estimate, which is the average squared residual, sums squared residuals across the entire sample and divides by $N - j$ with the N reflecting the entire sample ($M + F$) and j reflecting all of the coefficients in the pooled estimation, including the constant. Separate-sample estimation produces a separate estimate of s^2 for each subsample (e.g., s_m^2 and s_f^2). Each separate-sample estimation sums only the squared residuals from its subsample and divides only by the number of observations in its subsample, minus the number of coefficients in the subsample estimation, $N_s - j_s$. The subsample estimates are inefficient. In other words, we obtain better estimates of s^2 and, with them, of estimated variance-covariances of the coefficient estimates in pooled-sample than in separate-sample estimation—if, indeed, the residual variances are equal across subsamples.[5] In this case, the inefficiency manifests as one of the s_m^2 and s_f^2 being larger than it needs to be and the other smaller than it should be. More generally, some of the s_i^2 will be larger than they need to be and others smaller than they should be. To explore whether such a common error-variance assumption is warranted, we can test whether heteroskedasticity instead prevails. If the data insist that heteroskedasticity exists, then one can model that variance (or variance-covariance) structure and employ weighted (or feasible generalized) least squares in the pooled sample.

Other model restrictions, such as constraining some coefficients to be equal across subsamples while allowing others to vary, are also easier to implement in pooled-sample estimation and will also, if true, enhance coefficient and standard-error estimates' efficiency. For example, we may posit, or theory may establish, that some x affects males' and females' voting propensities equally (or equally and oppositely, or otherwise relatedly in some deterministic manner). In some contexts, accounting or other mathematical identities may even require certain relations between

---

5. In this case, the efficiency gains imply that estimated standard errors will be more accurate, not necessarily lower. As pooling borrows strength from the other subsamples to improve standard-error estimates, generally one (some) estimated effect(s) will be lower and (some) other(s) higher.

particular coefficients. Rather than estimate both of these effects separately, as separate-sample estimation all but requires,[6] one could in pooled-sample estimation simply refrain from including those dummy-variable interactions (or reverse the sign of those variables in the male or female sample, or analogously impose the constraints directly for other cases). As with a common-variance assumption, such cross-subsample restrictions can be tested, rather than assumed and imposed without testing, and again more conveniently in pooled-sample than in separate-subsample estimation. If the data insist that coefficients differ, this is easily allowed.

Thus, in short, compared to separate-sample estimation: (1) pooled-sample estimation can yield identical or superior interpretability; (2) it can encourage statistical comparison of effects over mere eyeballing; and (3) it may improve efficiency (precision) of estimation more easily if any efficiency-enhancing cross-subsample coefficient or error variance-covariance constraints are warranted. Therefore, if theory dictates that the effects of all variables should be dependent upon some x, we generally recommend that researchers present pooled-sample estimates as their final analysis—and report on the statistical certainty of any differences in effects they deem substantively important—even if they find conducting preliminary exploratory analysis in separate subsamples more convenient. We reiterate that while fully interactive pooled-sample estimation is preferable to separate-sample estimation, neither substitutes for a theoretically motivated model that identifies persuasively why the effect of some (set of) variable(s) should depend on x.[7]

Nonlinear Models

To this point, we have limited our discussion to interactive terms in linear models. However, we must also address interactions in nonlinear

---

6. To our knowledge, only some relatively complicated iterative procedure, like MCMC (Markov chain Monte Carlo), could succeed in imposing that some $\hat{\beta}_m = \hat{\beta}_f$ across separate-subsample estimations, for example, and correctly gauge the statistical uncertainty of that single coefficient estimate.

7. In multicategory cases, one can include $\mathbf{X}$, indicators for all the categories except one, and all the interactions of the former with the latter, in which case the excluded category becomes the suppressed reference group that serves as the baseline for comparison. Standard t-tests would in this case refer to whether the effect in the category in question differs significantly from that base case for that category's indicator. Alternatively, one could block-diagonalize $\mathbf{X}$, and then the coefficients would refer directly to the estimated effects of each x in each category, whereas tests of significance of any differences in estimated effects would require additional steps. In either case, one can interpret these interactive effects by calculating differences in predicted probabilities or derivatives (treating the derivatives of noncontinuous indicators as approximations).

models, which would include most models of qualitative dependent variables, given their prevalence in social science. For nonlinear models that include explicit linear-interactive terms among their right-hand-side variables, much of the discussion in preceding sections and chapters continues to apply. However, a further complication arises regarding the effect of x on y when right-hand-side variables are nonlinearly related to y by construction in the model. In logit or probit models of binary outcomes, for example, the effect of a variable x on y depends on the values of (all) the other variables z automatically due to the imposed nonlinear structure of the model. Thus, nonlinear models express conditional (i.e., interactive) relationships of the independent to dependent variables by construction, although they may also contain additional explicitly modeled linear interactions among the right-hand-side arguments of those nonlinear functions. The issue, then, arises regarding the proper interpretation of the effects of variables upon which a conditional relationship has been imposed, or assumed by construction, by virtue of the particular model specification employed.

Logit and probit models of dichotomous outcomes, for example, both (1) impose conditional relationships of x to y by construction and (2) use a sigmoidal (i.e., S-shaped) functional form implying specific character to those interactions. In these sigmoidal functional forms, the effects of changes in one variable on y are larger when the predicted probabilities are closer to the midpoint. Noting this, Nagler (1991), for example, critiques the claim of Wolfinger and Rosenstone (1980) that registration requirements discourage turnout to a greater extent among low education groups. He argues that this larger effect derives from the functional form assumed a priori and not necessarily from an explicit or direct interaction between education and registration requirements, for instance, that the less educated find surmounting registration requirements more difficult. The logit form by itself implies that education interacts with registration requirements and vice versa only because of and only through the other variable's effect on the overall propensity to vote. Insofar as being less educated puts one nearer a 0.5 probability of voting and being more educated puts one further from that point, registration requirements will have greater effect on the less educated's propensity to vote for that reason alone. Nagler tests whether education and registration requirements additionally interact explicitly to move a respondent along the S curve by including a specific linear interaction between education and registration requirement in the argument to a logit function. He also estimates logit coefficients on strict versus lax registration requirements separately in samples split by education (a

strategy discussed earlier in this chapter). He finds little support for
Wolfinger and Rosenstone's conclusion.[8]

The notion that multiple explicit interactions determine one's dependent variable suggests explicit modeling of those interactions, in as precise a fashion as theoretically possible. The defense for the specific form
of interactivity implicit in logit, probit, and related models is, in fact, explicit and theoretical in this way. First, the logit and probit functional
form implies a particular and very specific set of interactions to produce
S shapes. That such S shapes should describe the relations of independent
to dependent variables is substantively and theoretically derived from the
proposition that inducing probabilities to increase or to decrease becomes increasingly difficult, that is, requires larger movements in independent variables, as probabilities near one or zero (see also note 8). If
the researcher wishes to infer beyond the specific forms of interactions
that produce these S shapes, we concur with Nagler (1991) that he or she
must model those further interactions explicitly.

We now discuss in more detail the interpretation of effects in two
commonly used nonlinear models: probit and logit. For example, suppose some nonlinear function, $F(\cdot)$, often called a "link function," is used
to relate a binary outcome, y, with $\mathbf{x}'\boldsymbol{\beta}$, where y refers to a binary dependent variable, $\mathbf{x}'$ refers to a row vector of $k + 1$ regressors, and $\boldsymbol{\beta}$ refers

8. Similarly, Frant (1991) reviews the research of Berry and Berry (1990) on state lottery policy adoptions. Frant argues that Berry and Berry draw their conclusions about the
interaction between motivation, obstacles to innovation, and resources to overcome obstacles to innovation from the assumption inherent in the probit specification they employ.
Berry and Berry (1991), however, disagree. They believe that their theory suggests that they
estimate a probit model with no interactions or a linear probability model with a number
of multiplicative terms. However, they prefer the probit model because the complexly interactive theory driving their model would require "so large a number of multiplicative
terms as to render the model useless for empirical analysis because of extreme colinearity"
(578). To argue that the complexly interactive nature of one's theory debars explicit modeling of it is a very weak defense by itself for applying an arbitrary specific functional form
(probit) to allow all the independent variables to interact according to that specific functional form rather than explicitly to derive the form of these complex interactions from the
theory. As we suggest and Frant (1991) notes, a stronger argument in defense would have
been to demonstrate directly and explicitly that the theory implied specifically a set of interactions like those entailed inherently in a probit model, which indeed seems possible in
this case. To generalize the example to a form common in many contexts, an argument
might involve some overcoming of resistance from a broad set of conditions (explanatory
factors) being necessary to produce an outcome. It might also then invoke some notion of
a tipping point set by some values of this set of conditions and possibly even consider the
outcome to become increasingly "overdetermined" as the factors all push for the outcome. Such an argument, which seems similar to Berry and Berry's, would indeed imply
an S-shaped relation, such as logit or probit, between the explanatory factors and the outcome. Alternative sources or types of interactions, however, would not be inherent in sig-
moid functions lacking those further, explicit interactions.

to a column vector of coefficients. In such a case, one could model the probability that y takes the value one as $p(y = 1) \equiv p = F(\mathbf{x}'\boldsymbol{\beta})$.

In the probit case, $p = \Phi(\mathbf{x}'\boldsymbol{\beta})$, where Φ is the cumulative standard-normal distribution. Cumulative normal distributions are S shaped, and so ever larger increases or decreases in $\mathbf{x}'\boldsymbol{\beta}$ are required to increase or decrease the probability $y = 1$ as this probability draws closer to one or zero. In the logit case, $p = \Lambda(\mathbf{x}'\boldsymbol{\beta})$, where $\Lambda(\cdot)$ is the logit function: $\Lambda(\mathbf{x}'\boldsymbol{\beta}) = e^{\mathbf{x}'\boldsymbol{\beta}}/(1 + e^{\mathbf{x}'\boldsymbol{\beta}})$ or, equivalently, $\Lambda(\mathbf{x}'\boldsymbol{\beta}) = [1 + e^{-\mathbf{x}'\boldsymbol{\beta}}]^{-1}$. (Several other useful formulations of the logit function also exist.)

We begin with a simple probability model that omits explicit interaction terms:

$$p = F(\mathbf{x}'\boldsymbol{\beta}) = F(\beta_0 + \beta_x x + \beta_z z + \cdots + \beta_k w)$$

As always, the marginal effects of a variable x on p can be calculated by taking the first derivative of this function.[9] Note here the use of the chain rule in differentiating the function:

$$\partial p/\partial x = [\partial p/\partial F(\mathbf{x}'\boldsymbol{\beta})][\partial F(\mathbf{x}'\boldsymbol{\beta})/\partial x]$$

$$\partial p/\partial x = [\partial p/\partial F(\mathbf{x}'\boldsymbol{\beta})][\partial F(\mathbf{x}'\boldsymbol{\beta})/\partial x]$$

In the probit case, using the same model that omits explicit interaction terms, this is simply

$$p = \Phi(\mathbf{x}'\boldsymbol{\beta}) = \Phi(\beta_0 + \beta_x x + \beta_z z + \cdots + \beta_k w)$$

$$\equiv \int_{-\infty}^{\mathbf{x}'\boldsymbol{\beta}} \frac{1}{\sqrt{2\pi}} e^{-(1/2)t^2} \, dt \quad \text{where } t \equiv \beta_0 + \beta_x x + \beta_z z + \cdots + \beta_k w$$

$$\partial p/\partial x = [\partial \Phi(\mathbf{x}'\boldsymbol{\beta})/\partial x] = \phi(\mathbf{x}'\boldsymbol{\beta})\beta_x = \frac{1}{\sqrt{2\pi}} e^{-(1/2)(\mathbf{x}'\boldsymbol{\beta})^2} \times \beta_x$$

where $\phi(\mathbf{x}'\boldsymbol{\beta})$ is the standard-normal probability density function evaluated at $\mathbf{x}'\boldsymbol{\beta}$.[10] Thus, as is central to the theoretical proposition of an S-shaped relationship, the magnitude of effects of x on the probability that $y = 1$ is largest at $p = 0.5$ (at $\mathbf{x}'\boldsymbol{\beta} = 0$) and becomes smaller, approaching zero, as that probability goes to one or zero (as $\mathbf{x}'\boldsymbol{\beta}$ approaches in-

---

9. Note the distinction here between conceptualizing effects of a one-unit change in x literally computed (i.e., $\hat{p}_c - \hat{p}_a$) versus marginal effects, that is, effects of an infinitesimal change in x, $\partial y/\partial x$. Generally, the former is recommended for discrete variables and the latter for continuous variables. (See Greene 2003 for elaboration.)

10. The derivative of any cumulative probability distribution function (cdf), F, is the corresponding probability density function (pdf), f, and so the derivative of Φ, the cdf of the standard normal, is ϕ, the pdf of the standard normal.

finity or negative infinity). One sees also that the effect of each x depends on itself and all of the other variables, since all the covariates and their coefficients appear in the $\phi(\mathbf{x}'\boldsymbol{\beta})$ that multiplies the coefficient on x to determine the effect of x.

In the logit case, again for this model omitting explicit interaction terms, this is simply

$$p = \Lambda(\mathbf{x}'\boldsymbol{\beta}) = \left(1 + e^{-\mathbf{x}'\boldsymbol{\beta}}\right)^{-1} = \left(1 + e^{-(\beta_0 + \beta_x x + \beta_z z + \cdots + \beta_k w)}\right)^{-1}$$

$$\partial p / \partial x = \Lambda(\mathbf{x}'\boldsymbol{\beta})(1 - \Lambda(\mathbf{x}'\boldsymbol{\beta}))(\beta_x) \tag{42}$$

In the specific model of equation (42), this would be

$$\frac{\partial p}{\partial x} = \left(\frac{e^{\beta_0 + \beta_x x + \beta_z z + \cdots + \beta_k w}}{\left(1 + e^{\beta_0 + \beta_x x + \beta_z z + \cdots + \beta_k w}\right)^2} \right) \beta_x$$

$$= \Lambda(\beta_0 + \beta_x x + \beta_z z + \cdots + \beta_k w)$$

$$\times [1 - \Lambda(\beta_0 + \beta_x x + \beta_z z + \cdots + \beta_k w)]\beta_x$$

Obviously, as with probit, the effect of x depends on the values of x, $z, \ldots, w$ as well as the estimated coefficients for $\beta_0, \ldots, \beta_k$. We can also see, again as with probit, that the largest magnitude effects of x occur at $p = 0.5$, which occurs at $\mathbf{x}'\boldsymbol{\beta} = 0$, and that these effects become progressively smaller in magnitude as p approaches one or zero, which occurs as $\mathbf{x}'\boldsymbol{\beta}$ approaches positive or negative infinity, producing that familiar S shape again (although a slightly different S shape than probit produces).

When an explicit linear-interaction term (e.g., between x and z) is included in the $\mathbf{x}'\boldsymbol{\beta}$ part of the model, the effects of x continue to depend on the values of the other variables via the nonlinear form, specifically the S shape, of the model as before. In addition, movements along this S shape induced by movements in x depend directly on the value of z as well:

$$p = \Phi(\mathbf{x}'\boldsymbol{\beta}) = \int_{-\infty}^{\mathbf{x}'\boldsymbol{\beta}} \frac{1}{\sqrt{2\pi}} e^{-(1/2)t^2}\, dt \quad \text{where } t \equiv \begin{cases} \beta_0 + \beta_x x + \beta_z z \\ + \beta_{xz} xz + \cdots + \beta_k w \end{cases} \tag{43}$$

$$p = \Lambda(\mathbf{x}'\boldsymbol{\beta}) = \frac{e^{\beta_0 + \beta_x x + \beta_z z + \beta_{xz} xz + \cdots + \beta_k w}}{1 + e^{\beta_0 + \beta_x x + \beta_z z + \beta_{xz} xz + \cdots + \beta_k w}} \tag{44}$$

For illustration, we discuss a simple empirical example predicting turnout, using data from the 2004 National Election Studies. The dependent variable, *Voted*, is binary: 1 if the respondent voted; 0 if not. We model turnout as a function of two individual-level characteristics: education, ranging from one to seventeen years of *Schooling*, and strength of partisanship, *StrPID*, an ordinal measure equaling 0 for independents and 1 for leaning, 2 for weak, and 3 for strong partisans.[11] We interact education and strength of partisanship to explore whether education explicitly conditions the effect of strength of partisanship and vice versa. A researcher might argue that education and strength of partisanship each bring resources and motivation that reinforce each other in reducing the costs or increasing the benefits of voting, such that increases in one variable will boost the impact of the other in generating the net benefit to the individual of voting that relates nonlinearly (specifically: sigmoidally) to that individual's propensity to vote. Alternatively, the researcher might suspect the opposite: that educational and partisan resources and motivations undermine each other, such that increases in one variable contribute less to the net benefit of voting when the other is high than when it is low. Notice how these propositions argue something further than that the effect of one variable is higher or lower when the other is lower or higher *because both augment (detriment) the propensity to vote and so each has less effect when the other already leans the individual far toward or away from voting.* This last possibility is what the S-shaped function relating education and partisanship to vote propensity already assumes. Formally, we specify the following model (a fully specified model of turnout would, of course, include several additional covariates):

$$Voted = F(\beta_0 + \beta_{Sch}Schooling + \beta_{Str}StrPID + \beta_{Sch \times Str}Schooling$$
$$\times StrPID + \varepsilon)$$

The logit and probit estimates appear in table 27.

The effects of x can be calculated using the derivative method or the method of differences in predicted probabilities. For the first-derivative approach, interpretation of a model with an explicit interaction in addition to its implicit ones would again require application of the chain rule. For logit:

$$\frac{\partial p}{\partial x} = \frac{\partial p}{\partial \mathbf{x}'\boldsymbol{\beta}} \frac{\partial \mathbf{x}'\boldsymbol{\beta}}{\partial x} = -(1 + e^{-\mathbf{x}'\boldsymbol{\beta}})^{-2} e^{-\mathbf{x}'\boldsymbol{\beta}}(-1)\frac{\partial \mathbf{x}'\boldsymbol{\beta}}{\partial x}$$

$$= \left(\frac{e^{\mathbf{x}'\boldsymbol{\beta}}}{1 + e^{\mathbf{x}'\boldsymbol{\beta}}}\right)\left(\frac{1}{1 + e^{\mathbf{x}'\boldsymbol{\beta}}}\right)(\beta_x + \beta_{xz}z)$$

$$= [\Lambda(\mathbf{x}'\boldsymbol{\beta})][1 - \Lambda(\mathbf{x}'\boldsymbol{\beta})]\frac{\partial \mathbf{x}'\boldsymbol{\beta}}{\partial x} = p(1 - p)(\beta_x + \beta_{xz}z) \qquad (45)$$

---

11. Here, as is common in such cases, we are treating the ordinal information on partisan leanings recorded by this measure as interval (or effectively interval, plus only some unimportant and unproblematic noise) by giving it simple linear coefficients in $\mathbf{x}'\boldsymbol{\beta}$.

This is the same expression as before except that now the effect of x depends not only on the other $\mathbf{x}$ through $[\Lambda(\mathbf{x'\beta})][1 - \Lambda(\mathbf{x'\beta})]$ but also and again on the value of z in the manner implied by the linear interaction of x and z contained in $\mathbf{x}$. Thus, z modifies the effect of x on p not only by its role in the calculation of $\Lambda(\mathbf{x'\beta})$, where it enters in the $+\beta_z z + \beta_{xz} xz$ terms, but also in the final term, $\partial \mathbf{x'\beta}/\partial x$, where it enters in the expression $\partial \mathbf{x'\beta}/\partial x = \beta_x + \beta_{xz} z$. The former role is that imposed by the assumed sigmoidal relationship from independent to dependent variables; the latter role is imposed by the explicit interaction term as z conditions the effect of x on movement along that S shape.

Similarly, for the probit model, when there is an explicit interaction between x and z:

$$\frac{\partial p}{\partial x} = \frac{\partial p}{\partial \mathbf{x'\beta}} \frac{\partial \mathbf{x'\beta}}{\partial x} = \phi(\mathbf{x'\beta})\frac{\partial \mathbf{x'\beta}}{\partial x} = \phi(\mathbf{x'\beta})(\beta_x + \beta_{xz} z)$$

$$= \frac{1}{\sqrt{2\pi}} e^{-(\frac{1}{2})(\mathbf{x'\beta})^2} \times (\beta_x + \beta_{xz} z) \tag{46}$$

In our example, the marginal effects of *Schooling* would be calculated at specific values of *Schooling* along varying values of *StrPID*, given as $\partial \hat{p}/\partial x = \hat{p}(1 - \hat{p})(\hat{\beta}_x + \hat{\beta}_{xz} z)$ in the logit case and $\partial \hat{p}/\partial x =$

TABLE 27. Logit and Probit Regression Results, *Turnout*

	Logit Estimates Coefficient (standard error) *p*-Value	Probit Estimates Coefficient (standard error) *p*-Value
Years of Schooling	0.310 (0.065) 0.000	0.190 (0.037) 0.000
Strength of Partisanship	0.904 (0.445) 0.042	0.607 (0.251) 0.015
Years of Schooling × *Strength of Partisanship*	−0.021 (0.034) 0.536	−0.019 (0.019) 0.321
Intercept	−3.842 (0.852) 0.000	−2.340 (0.489) 0.000
$N(df)$	1,065 (1,062)	1,065 (1,062)
lnL	−476.26	−476.04
$P > \chi^2$	0.000	0.000

Note: Cell entries are the estimated coefficient, with standard error in parentheses, and two-sided *p*-level (probability $|T| > t$) referring to the null hypothesis that $\beta = 0$ in italics.

$\phi(\mathbf{x}'\hat{\boldsymbol{\beta}})(\hat{\beta}_x + \hat{\beta}_{xz}z)$ in the probit case. Table 28 and table 29 provide the marginal effects of *Schooling* and *StrPID*, respectively, holding *Schooling* and *StrPID* at substantively interesting values. A sample calculation of the marginal effect of *Schooling*, when *Schooling* = 12 and *StrPID* = 3, using the logit results, is

$$\frac{\partial \hat{p}}{\partial Sch} = \left(\frac{e^{-3.84+0.31\times12+0.904\times3-0.021\times12\times3}}{1+e^{-3.84+0.31\times12+0.904\times3-0.021\times12\times3}}\right)\left(1-\frac{e^{-3.84+0.31\times12+0.904\times3-0.021\times12\times3}}{1+e^{-3.84+0.31\times12+0.904\times3-0.021\times12\times3}}\right)$$

$$\times\,(0.31 + -0.021 \times 3)$$

$$=(0.861)(1 - 0.861)(0.31 + -0.021 \times 3) \approx 0.029$$

Alternatively, one could calculate the predicted probabilities, $\hat{p}$, with appropriate confidence intervals. The intuition behind calculating the predicted probabilities in a nonlinear model is exactly the same as that behind calculating predicted values of y in a linear model. The nonlinear model merely requires an additional step, in projecting the linear index (i.e., the sum of the coefficients times their covariates) through the non-linear model onto probability space (in the cases of logit and probit). For example, suppose we estimated the following relationship:

$$p = F(\beta_0 + \beta_x x + \beta_z z + \beta_{xz}xz)$$

Denote the predicted probabilities $\hat{F} = F(\mathbf{x}'\hat{\boldsymbol{\beta}})$, with the linear index, $\mathbf{x}'\hat{\boldsymbol{\beta}}$, computed in identical fashion to the linear-regression case:

$$\mathbf{x}'\hat{\boldsymbol{\beta}} = \hat{\beta}_0 + \hat{\beta}_x x + \hat{\beta}_z z + \hat{\beta}_{xz}xz$$

TABLE 28. Marginal Effects of *Schooling*, Using Logit Results

	Independents	Leaning Partisans	Weak Partisans	Strong Partisans
Years of Schooling = 9	0.059 (0.007)	0.070 (0.008)	0.065 (0.010)	0.046 (0.016)
Years of Schooling = 12	0.077 (0.016)	0.067 (0.010)	0.048 (0.008)	0.029 (0.008)
Years of Schooling = 15	0.066 (0.011)	0.046 (0.005)	0.029 (0.003)	0.016 (0.003)

Note: Cell entries are the estimated marginal effect, with standard error in parentheses.

TABLE 29. Marginal Effects of *Strength of Partisanship*, Using Logit Results

	Independents	Leaning Partisans	Weak Partisans	Strong Partisans
Years of Schooling = 9	0.137 (0.017)	0.173 (0.035)	0.172 (0.035)	0.134 (0.018)
Years of Schooling = 12	0.162 (0.020)	0.152 (0.022)	0.117 (0.014)	0.078 (0.006)
Years of Schooling = 15	0.125 (0.032)	0.093 (0.022)	0.063 (0.011)	0.039 (0.004)

Note: Cell entries are the estimated marginal effect, with standard error in parentheses.

After calculation of the linear index, the researcher must use the link function, $F(\mathbf{x}'\hat{\boldsymbol{\beta}})$ (here, the logit $\Lambda(\mathbf{x}'\hat{\boldsymbol{\beta}})$ or probit $\Phi(\mathbf{x}'\hat{\boldsymbol{\beta}})$), to convert the linear index into probability space. In either case, the predicted probabilities would be calculated at various values of x (say, between x_a and x_c), holding z at some substantively meaningful and logically relevant value (e.g., its sample mean, $\bar{z}$) and of course allowing xz to vary from $x_a\bar{z}$ to $x_c\bar{z}$.

Thus, to calculate the effect on the predicted probability of a discrete change in x, say, from x_a and x_c, one would simply first compute the linear index at x_a and x_c:

$$(\mathbf{x}'\hat{\boldsymbol{\beta}})_a = \hat{\beta}_0 + \hat{\beta}_x x_a + \hat{\beta}_z \bar{z} + \hat{\beta}_{xz} x_a \bar{z};$$

$$(\mathbf{x}'\hat{\boldsymbol{\beta}})_c = \hat{\beta}_0 + \hat{\beta}_x x_c + \hat{\beta}_z \bar{z} + \hat{\beta}_{xz} x_c \bar{z}$$

Then one would project each linear index into probability space; for the logit case:

$$\hat{p}_a = \frac{e^{(\mathbf{x}'\hat{\boldsymbol{\beta}})_a}}{1 + e^{(\mathbf{x}'\hat{\boldsymbol{\beta}})_a}}; \qquad \hat{p}_c = \frac{e^{(\mathbf{x}'\hat{\boldsymbol{\beta}})_c}}{1 + e^{(\mathbf{x}'\hat{\boldsymbol{\beta}})_c}}$$

And then one simply computes the difference between the two probabilities: $\hat{p}_c - \hat{p}_a$. For probit, the process is identical except that one uses $\Phi(\mathbf{x}'\hat{\boldsymbol{\beta}})_a$ instead of $[1 + e^{-(\mathbf{x}'\hat{\boldsymbol{\beta}})_a}]^{-1}$, that is, the cumulative standard normal rather than the logit, as the link function.

We reiterate our strong recommendation that researchers compute and report measures of uncertainty around marginal effects and predicted probabilities. Standard errors for marginal effects can be computed by the delta method, as described in most statistics texts, for example, Greene (2003, 70).[12] The variance of any nonlinear function of parameter estimates, such as an estimated marginal effect like $\partial \hat{p}/\partial x$, is approximated asymptotically as a linear function of the estimated variance-covariance matrix of the parameter estimates, here $\widehat{V(\hat{\boldsymbol{\beta}})}$, and the derivative of the function with respect to $\hat{\boldsymbol{\beta}}$:[13]

---

12. For confidence intervals around predicted levels, $\hat{p}$, a simpler expedient of calculating confidence intervals for the linear $\mathbf{x}'\hat{\boldsymbol{\beta}}$ and then translating those bounds to probability space using the link function will also suffice and, indeed, would have the advantage of constraining the confidence interval bounds to lie between zero and one, which the delta method's linearization strategy would not. That expedient would seem unavailable for confidence intervals around marginal effects and differences, however.

13. The derivative of a function with respect to a vector of its arguments is called a gradient and denoted $\nabla_{\hat{\boldsymbol{\beta}}}$, but we eschew this terminology and notation as probably less familiar to many readers.

$$\widehat{V\left(\frac{\partial \hat{p}}{\partial x}\right)} \approx \left[\frac{\partial\left(\frac{\partial \hat{p}}{\partial x}\right)}{\partial \hat{\boldsymbol{\beta}}'}\right]\left[\widehat{V(\hat{\boldsymbol{\beta}})}\right]\left[\frac{\partial\left(\frac{\partial \hat{p}}{\partial x}\right)}{\partial \hat{\boldsymbol{\beta}}'}\right]' \tag{47}$$

We now apply this to the logit case, where $\hat{p} = (1 + e^{-\mathbf{x}'\hat{\boldsymbol{\beta}}})^{-1}$, $\partial\hat{p}/\partial x = \hat{p}(1 - \hat{p}) \, \partial\mathbf{x}'\hat{\boldsymbol{\beta}}/\partial x$.[14]

Next, using the product rule[15] to solve $[\partial(\partial\hat{p}/\partial x)/\partial\hat{\boldsymbol{\beta}}']$:

$$\partial(\partial\hat{p}/\partial x)/\partial\hat{\boldsymbol{\beta}}' = [\partial(\partial\mathbf{x}'\hat{\boldsymbol{\beta}}/\partial x)/\partial\hat{\boldsymbol{\beta}}'](\hat{p}(1 - \hat{p})) + [\partial\hat{p}/\partial\hat{\boldsymbol{\beta}}']$$
$$\times ((1 - \hat{p})(\partial\mathbf{x}'\hat{\boldsymbol{\beta}}/\partial x)) + [\partial(1 - \hat{p})/\partial\hat{\boldsymbol{\beta}}'](\hat{p}(\partial\mathbf{x}'\hat{\boldsymbol{\beta}}/\partial x))$$

Reexpressing terms, given that $\partial\mathbf{x}'\hat{\boldsymbol{\beta}}/\partial x = \hat{\beta}_x + \hat{\beta}_{xz}z$: $\partial(\partial\mathbf{x}'\hat{\boldsymbol{\beta}}/\partial x)/\partial\hat{\boldsymbol{\beta}}' = \partial(\hat{\beta}_x + \hat{\beta}_{xz}z)/\partial\hat{\boldsymbol{\beta}}'$. Let $\hat{\beta}_x + \hat{\beta}_{xz}z = \mathbf{r}'\hat{\boldsymbol{\beta}}$, where $\mathbf{r}' = [1 \; 0 \; z \; 0]$, assuming the estimated coefficients are arranged as $\hat{\boldsymbol{\beta}}' = [\hat{\beta}_x \; \hat{\beta}_z \; \hat{\beta}_{xz} \; \hat{\beta}_0]$, in that order. Differentiating: $\partial(\partial\mathbf{x}'\hat{\boldsymbol{\beta}}/\partial x)/\partial\hat{\boldsymbol{\beta}}' = \partial\mathbf{r}'\hat{\boldsymbol{\beta}}/\partial\hat{\boldsymbol{\beta}}' = \mathbf{r}'$.

For the next term, $\partial\hat{p}/\partial\hat{\boldsymbol{\beta}}'$:

$$\partial\hat{p}/\partial\hat{\boldsymbol{\beta}}' = (\partial\hat{p}/\partial(\mathbf{x}'\hat{\boldsymbol{\beta}}))(\partial(\mathbf{x}'\hat{\boldsymbol{\beta}})/\partial\hat{\boldsymbol{\beta}}') = \frac{e^{-\mathbf{x}'\boldsymbol{\beta}}}{(1 + e^{-\mathbf{x}'\boldsymbol{\beta}})^2}\mathbf{x}' = \hat{p}(1 - \hat{p})\mathbf{x}'$$

And for the next term, $\partial(1 - \hat{p})/\partial\hat{\boldsymbol{\beta}}'$:

$$\frac{\partial(1 - \hat{p})}{\partial\hat{\boldsymbol{\beta}}'} = \frac{\partial(1 - \hat{p})}{\partial(\mathbf{x}'\hat{\boldsymbol{\beta}})}\frac{\partial(\mathbf{x}'\hat{\boldsymbol{\beta}})}{\partial\hat{\boldsymbol{\beta}}'} = \frac{\partial\left(1 - (1 + e^{-\mathbf{x}'\boldsymbol{\beta}})^{-1}\right)}{\partial(\mathbf{x}'\hat{\boldsymbol{\beta}})}\frac{\partial(\mathbf{x}'\hat{\boldsymbol{\beta}})}{\partial\hat{\boldsymbol{\beta}}'}$$

$$= \frac{-e^{-\mathbf{x}'\boldsymbol{\beta}}}{(1 + e^{-\mathbf{x}'\boldsymbol{\beta}})^2}\mathbf{x}' = -\left(\hat{p}(1 - \hat{p})\mathbf{x}'\right)$$

Substituting:

$$\partial(\partial\hat{p}/\partial x)/\partial\hat{\boldsymbol{\beta}}' = \mathbf{r}'(\hat{p}(1 - \hat{p})) + [\hat{p}(1 - \hat{p})\mathbf{x}']((1 - \hat{p})(\partial\mathbf{x}'\boldsymbol{\beta}/\partial x))$$
$$+ [-(\hat{p}(1 - \hat{p})\mathbf{x}')](\hat{p}(\partial\mathbf{x}'\boldsymbol{\beta}/\partial x))$$

Substituting $\partial\mathbf{x}'\hat{\boldsymbol{\beta}}/\partial x = \hat{\beta}_x + \hat{\beta}_{xz}z$:

$$\partial(\partial\hat{p}/\partial x)/\partial\hat{\boldsymbol{\beta}}' = (\hat{p}(1 - \hat{p}))(\mathbf{r}' + (1 - 2\hat{p})(\hat{\beta}_x + \hat{\beta}_{xz}z)\mathbf{x}')$$

Then substituting into equation (47):

$$\widehat{V(\partial\hat{p}/\partial x)} \approx (\hat{p}(1 - \hat{p}))((\mathbf{r}' + (1 - 2\hat{p})(\hat{\beta}_x + \hat{\beta}_{xz}z)\mathbf{x}'))\widehat{V(\hat{\boldsymbol{\beta}})}(\hat{p}(1 - \hat{p}))$$
$$\times ((\mathbf{r}' + (1 - 2\hat{p})(\hat{\beta}_x + \hat{\beta}_{xz}z)\mathbf{x}'))'$$
$$= (\hat{p}(1 - \hat{p}))^2(\mathbf{r}' + (1 - 2\hat{p})(\hat{\beta}_x + \hat{\beta}_{xz}z)\mathbf{x}')\widehat{V(\hat{\boldsymbol{\beta}})}$$
$$\times (\mathbf{r} + (1 - 2\hat{p})(\hat{\beta}_x + \hat{\beta}_{xz}z)\mathbf{x})$$

---

14. In the simple case that contains no explicit interaction, $\partial\hat{p}/\partial x = \hat{p}(1 - \hat{p}) \, (\partial\mathbf{x}'\hat{\boldsymbol{\beta}}/\partial x)$ $= \hat{p}(1 - \hat{p})\hat{\beta}_x$. When x interacts with another variable, z, as in equation (44), then $\partial\hat{p}/\partial x = \hat{p}(1 - \hat{p})(\hat{\beta}_x + \hat{\beta}_{xz}z)$.

15. Recall that $\partial(f(x)g(x))/\partial x = \partial f(x)/\partial x \, g(x) + \partial g(x)/\partial x \, f(x)$.

Note that $(1 - 2\hat{p})(\hat{\beta}_x + \hat{\beta}_{xz}z)$ is a scalar for a given set of values of x, z, and xz that scales the values in vector $\mathbf{x}'$. Let s_L be the value of the scaling value in the logit: $s_L = (1 - 2\hat{p})(\hat{\beta}_x + \hat{\beta}_{xz}z)$:

$$\widehat{V(\partial\hat{p}/\partial x)} \approx (\hat{p}(1 - \hat{p}))^2(\mathbf{r}' + s_L\mathbf{x}')\widehat{V(\hat{\beta})}(\mathbf{r} + s_L\mathbf{x})$$

$$= (\hat{p}(1 - \hat{p}))^2\left(\mathbf{r}'\widehat{V(\hat{\beta})}\mathbf{r} + 2s_L\mathbf{x}'\widehat{V(\hat{\beta})}\mathbf{r} + s_L^2\mathbf{x}'\widehat{V(\hat{\beta})}\mathbf{x}\right)$$

Using our empirical example, we can calculate the estimated variance around the estimated marginal effect of *Schooling*, when *Schooling* = 12 and *StrPID* = 3. In this example, $\mathbf{x}' = [12\ 3\ 36\ 1]$; the value at which *Schooling* is held is located in the first column; the value at which *StrPID* is held is in the second column; the interaction term's value appears in the third column; and a 1 is located in the last column, to represent the intercept. We established previously that $(\hat{p} \mid Sch = 12, Str = 3) = 0.861$. Because we are taking $\partial\hat{p}/\partial x$ with respect to *Schooling*, and because the value of *StrPID* is 3, $\mathbf{r}' = [1\ 0\ 3\ 0]$. As with linear regression, the estimated variance-covariance matrix of the estimated logit or probit coefficients can be easily called by a postestimation command. In this case,

$$\mathbf{V}(\hat{\boldsymbol{\beta}}) = \begin{bmatrix} 0.004 & 0.024 & -0.002 & -0.055 \\ 0.024 & 0.198 & -0.015 & -0.323 \\ -0.002 & -0.015 & 0.001 & 0.024 \\ -0.055 & -0.323 & 0.024 & 0.726 \end{bmatrix},$$

a 4×4 matrix that lists the estimated coefficient variances and covariances in the order in which they appear in the regression results and corresponding with the order in which values are arrayed in $\mathbf{x}'$. Substituting the set values in $\mathbf{x}'$, the values in $\mathbf{r}'$, and the estimated values for $\hat{p}$ and $\widehat{V(\hat{\boldsymbol{\beta}})}$:

$$\widehat{V(\partial\hat{p}/\partial x)} \approx (0.861(1 - 0.861))^2$$

$$\left([1\ 0\ 3\ 0] \begin{bmatrix} 0.004 & 0.024 & -0.002 & -0.055 \\ 0.024 & 0.198 & -0.015 & -0.323 \\ -0.002 & -0.015 & 0.001 & 0.024 \\ -0.055 & -0.323 & 0.024 & 0.726 \end{bmatrix} \begin{bmatrix} 1 \\ 0 \\ 3 \\ 0 \end{bmatrix} \right.$$

$$\times\ +2s_L[12\ 3\ 36\ 1] \begin{bmatrix} 0.004 & 0.024 & -0.002 & -0.055 \\ 0.024 & 0.198 & -0.015 & -0.323 \\ -0.002 & -0.015 & 0.001 & 0.024 \\ -0.055 & -0.323 & 0.024 & 0.726 \end{bmatrix} \begin{bmatrix} 1 \\ 0 \\ 3 \\ 0 \end{bmatrix}$$

$$\left. +s_L^2[12\ 3\ 36\ 1] \begin{bmatrix} 0.004 & 0.024 & -0.002 & -0.055 \\ 0.024 & 0.198 & -0.015 & -0.323 \\ -0.002 & -0.015 & 0.001 & 0.024 \\ -0.055 & -0.323 & 0.024 & 0.726 \end{bmatrix} \begin{bmatrix} 12 \\ 3 \\ 36 \\ 1 \end{bmatrix} \right)$$

where $s_L = (1 - 2 \times 0.861)(0.31 - 0.021 \times 3)$. A standard statistical package or a spreadsheet program can easily perform these calculations.

Similarly for the probit case, standard errors around marginal effects are calculated following equation (47); specifying $\hat{p} = \Phi(\mathbf{x}'\hat{\boldsymbol{\beta}})$, we have $\partial\hat{p}/\partial x = \phi(\mathbf{x}'\hat{\boldsymbol{\beta}})\partial\mathbf{x}'\hat{\boldsymbol{\beta}}/\partial x$. Using the product rule, $\partial(\partial\hat{p}/\partial x)/\partial\hat{\boldsymbol{\beta}}' = [\partial(\partial\mathbf{x}'\hat{\boldsymbol{\beta}}/\partial x)/\partial\hat{\boldsymbol{\beta}}']\phi(\mathbf{x}'\hat{\boldsymbol{\beta}}) + (\partial\mathbf{x}'\hat{\boldsymbol{\beta}}/\partial x)[\partial\phi(\mathbf{x}'\hat{\boldsymbol{\beta}})/\partial\hat{\boldsymbol{\beta}}']$. Reexpressing the first term in brackets: $\partial(\partial\mathbf{x}'\hat{\boldsymbol{\beta}}/\partial x)/\partial\hat{\boldsymbol{\beta}}' = \mathbf{r}'$. The second term in brackets is $\partial\phi(\mathbf{x}'\hat{\boldsymbol{\beta}})/\partial\hat{\boldsymbol{\beta}}' = (\partial\phi(\mathbf{x}'\hat{\boldsymbol{\beta}})/\partial(\mathbf{x}'\hat{\boldsymbol{\beta}}))(\partial(\mathbf{x}'\hat{\boldsymbol{\beta}})/\partial\hat{\boldsymbol{\beta}}') = (1/\sqrt{2\pi}\ e^{-(\frac{1}{2})(\mathbf{x}'\hat{\boldsymbol{\beta}})^2})(-\mathbf{x}'\hat{\boldsymbol{\beta}})(\partial(\mathbf{x}'\hat{\boldsymbol{\beta}})/\partial\hat{\boldsymbol{\beta}}') = -(\phi(\mathbf{x}'\hat{\boldsymbol{\beta}}))(\mathbf{x}'\hat{\boldsymbol{\beta}})\mathbf{x}'$. Substituting into $\partial(\partial\hat{p}/\partial x)/\partial\hat{\boldsymbol{\beta}}'$: $\partial(\partial\hat{p}/\partial x)/\partial\hat{\boldsymbol{\beta}}' = [\mathbf{r}']\phi(\mathbf{x}'\hat{\boldsymbol{\beta}}) - (\partial\mathbf{x}'\hat{\boldsymbol{\beta}}/\partial x)(\phi(\mathbf{x}'\hat{\boldsymbol{\beta}}))(\mathbf{x}'\hat{\boldsymbol{\beta}})\mathbf{x}' = \phi(\mathbf{x}'\hat{\boldsymbol{\beta}})(\mathbf{r}' - (\hat{\beta}_x + \hat{\beta}_{xz}z)(\mathbf{x}'\hat{\boldsymbol{\beta}})\mathbf{x}')$. Then substituting into equation (47):

$$\widehat{V(\partial\hat{p}/\partial x)} \approx [(\phi(\mathbf{x}'\hat{\boldsymbol{\beta}}))(\mathbf{r}' - \mathbf{x}'\hat{\boldsymbol{\beta}}(\hat{\beta}_x + \hat{\beta}_{xz}z)\mathbf{x}')]\widehat{V(\hat{\boldsymbol{\beta}})}[(\phi(\mathbf{x}'\hat{\boldsymbol{\beta}}))$$

$$\times (\mathbf{r}' - \mathbf{x}'\hat{\boldsymbol{\beta}}(\hat{\beta}_x + \hat{\beta}_{xz}z)\mathbf{x}')]'$$

$$= (\phi(\mathbf{x}'\hat{\boldsymbol{\beta}}))^2 (\mathbf{r}' - \mathbf{x}'\hat{\boldsymbol{\beta}}(\hat{\beta}_x + \hat{\beta}_{xz}z)\mathbf{x}')\widehat{V(\hat{\boldsymbol{\beta}})}[(\mathbf{r} - \hat{\boldsymbol{\beta}}'\mathbf{x}(\hat{\beta}_x + \hat{\beta}_{xz}z)\mathbf{x})]$$

Again, note that $\mathbf{x}'\hat{\boldsymbol{\beta}}(\hat{\beta}_x + \hat{\beta}_{xz}z)$ is a scalar for a given set of values of x, z, and xz. Let $s_P = \mathbf{x}'\hat{\boldsymbol{\beta}}(\hat{\beta}_x + \hat{\beta}_{xz}z)$. Substituting:

$$\widehat{V(\partial\hat{p}/\partial x)} \approx (\phi(\mathbf{x}'\hat{\boldsymbol{\beta}}))^2 (\mathbf{r}' - s_P\mathbf{x}')\widehat{V(\hat{\boldsymbol{\beta}})}(\mathbf{r} - s_P\mathbf{x})$$

$$= (\phi(\mathbf{x}'\hat{\boldsymbol{\beta}}))^2 (\mathbf{r}'\widehat{V(\hat{\boldsymbol{\beta}})}\mathbf{r} - 2s_P\mathbf{x}'\widehat{V(\hat{\boldsymbol{\beta}})}\mathbf{r} + s_P^2\mathbf{x}'\widehat{V(\hat{\boldsymbol{\beta}})}\mathbf{x})$$

For standard errors around predicted probabilities, we can also use the delta method. In the logit case, $\widehat{V(\hat{p})} \approx [\partial\hat{p}/\partial\hat{\boldsymbol{\beta}}]' [\widehat{V(\hat{\boldsymbol{\beta}})}][\partial\hat{p}/\partial\hat{\boldsymbol{\beta}}] = [\hat{p}(1 - \hat{p})\mathbf{x}'][\widehat{V(\hat{\boldsymbol{\beta}})}][\hat{p}(1 - \hat{p})\mathbf{x}] = (\hat{p}(1 - \hat{p}))^2 \mathbf{x}'[\widehat{V(\hat{\boldsymbol{\beta}})}]\mathbf{x}$. That is, square $\hat{p}(1 - \hat{p})$ and multiply the result by the estimated variance-covariance matrix of the estimated coefficients, pre- and postmultiplied by the $\mathbf{x}$ vector specified at the values of interest. In the probit case, $\widehat{V(\hat{p})} \approx [-(\phi(\mathbf{x}'\hat{\boldsymbol{\beta}}))\mathbf{x}]'[\widehat{V(\hat{\boldsymbol{\beta}})}][- (\phi(\mathbf{x}'\hat{\boldsymbol{\beta}}))\mathbf{x}] = (\phi(\mathbf{x}'\hat{\boldsymbol{\beta}}))^2 \mathbf{x}'[\widehat{V(\hat{\boldsymbol{\beta}})}]\mathbf{x}$. As with linear-regression models, predicted probabilities are most effective presentationally when graphed with confidence intervals. Confidence intervals can be generated using the same formulas: $\hat{p} \pm t_{df,p} \sqrt{\widehat{V(\hat{p})}}$.

Calculation of the standard error for the difference between two predicted probabilities, say, those reflecting the effect of a specific change in x from x_a to x_c, follows the same delta method:

$$\widehat{V(\hat{F}_c - \hat{F}_a)} \approx \left[\frac{\partial(\hat{F}_c - \hat{F}_a)}{\partial\hat{\boldsymbol{\beta}}}\right]' \left[\widehat{V(\hat{\boldsymbol{\beta}})}\right] \left[\frac{\partial(\hat{F}_c - \hat{F}_a)}{\partial\hat{\boldsymbol{\beta}}}\right] = \left[\frac{\partial\hat{F}_c}{\partial\hat{\boldsymbol{\beta}}} - \frac{\partial\hat{F}_a}{\partial\hat{\boldsymbol{\beta}}}\right]' \left[\widehat{V(\hat{\boldsymbol{\beta}})}\right]$$

$$\times \left[\frac{\partial\hat{F}_c}{\partial\hat{\boldsymbol{\beta}}} - \frac{\partial\hat{F}_a}{\partial\hat{\boldsymbol{\beta}}}\right]$$

$$= [\hat{f}_c\mathbf{x}'_c - \hat{f}_a\mathbf{x}'_a][\widehat{V(\hat{\boldsymbol{\beta}})}][\hat{f}_c\mathbf{x}_c - \hat{f}_a\mathbf{x}_a]$$

Here $\hat{F}_a$ and $\hat{F}_c$ are the link function (logit or probit here), and $\hat{f}_a$ and $\hat{f}_c$ are their derivatives with respect to $\mathbf{x}'\hat{\boldsymbol{\beta}}$, ($\hat{p}(1 - \hat{p})$ for logit and $\phi(\mathbf{x}'\hat{\boldsymbol{\beta}})$ for probit). These link functions and derivatives are evaluated at $\mathbf{x}_a$ and $\mathbf{x}_c$, respectively.

Many existing statistical software packages will calculate these standard errors of estimated probabilities for the researcher, and some will even calculate standard errors for derivatives or differences at user-given levels of the variables. Our intention here is to reemphasize the importance of examining effects rather than simply coefficients (or predicted levels), be they estimated in a linear or nonlinear specification, and to provide readers with a sense of the mathematics underlying the calculation of these estimated effects and their corresponding standard errors.

Random-Effects Models and Hierarchical Models

When modeling relationships between a set of covariates, $\mathbf{X}$, and a dependent variable, y, scholars make assumptions about the deterministic (i.e., fixed) versus stochastic (i.e., random) nature of those relationships. In the interaction context, for example, scholars might propose that the effects of x and of z on y depend either deterministically or stochastically on the other variable. The burgeoning "random-effects" literature proposes the latter, probabilistic, relationship. (The related multilevel-model or hierarchical-model literature addresses a similar issue, although possibly with different assumptions about the properties of the stochastic aspects of the relationships: see the discussion that follows.)[16]

Let us start thus:

$$y = \beta_0 + \beta_1 x + \beta_2 z + \varepsilon \tag{48}$$

As before, the linear-interactive specification of the posited interactive relationships could be

$$\beta_0 = \gamma_0 + \gamma_1 x + \gamma_2 z, \qquad \beta_1 = \delta_1 + \delta_2 z, \qquad \text{and} \quad \beta_2 = \delta_3 + \delta_4 x \tag{49}$$

---

16. For more thorough discussion of the issues in this section, see Franzese (2005).

in the deterministic case, suggesting our standard linear-interactive regression model:

$$y = \gamma_0 + \beta_x x + \beta_z z + \beta_{xz} xz + \varepsilon \tag{50}$$

where $\beta_x = \gamma_1 + \delta_1$, $\beta_z = \gamma_2 + \delta_3$, $\beta_{xz} = \delta_2 + \delta_4$. Notice, however, that this standard model in fact assumes that the effect of x on y varies with z, and the effect of z on y varies with x, *without error*. Likewise, the intercept does not vary across repeated samples. A linear-interactive model with random effects would instead be

$$\beta_0 = \gamma_0 + \gamma_1 x + \gamma_2 z + \varepsilon_0, \qquad \beta_1 = \delta_1 + \delta_2 z + \varepsilon_1, \qquad \text{and}$$

$$\beta_2 = \delta_3 + \delta_4 x + \varepsilon_2 \tag{51}$$

suggesting the following similar-looking linear-interactive regression model:

$$y = \gamma_0 + \beta_x x + \beta_z z + \beta_{xz} xz + \varepsilon^* \tag{52}$$

but with $\varepsilon^* = \varepsilon + \varepsilon_0 + \varepsilon_1 x + \varepsilon_2 z$.

Thus, the distinction between the deterministically interactive and the stochastically interactive models occurs only in the "error" term; the two models are identical except for the difference between ε and ε^*. In the first case, where the conditioning effects are assumed to be deterministic, OLS would be BLUE, that is, yielding the best (most efficient), linear unbiased estimates (provided the model is also correctly specified in other regards, of course). In the latter case, where effects are assumed stochastic, or probabilistic, one suspects that OLS estimates might not be BLUE. Notice, however, that, assuming all the stochastic terms have mean zero, $E(\varepsilon, \varepsilon_0, \varepsilon_1, \varepsilon_2) = 0$, and do not covary with the regressors, $C(\{\varepsilon, \varepsilon_0, \varepsilon_1, \varepsilon_2\}, \mathbf{x}) = 0$, as commonly done in most regression contexts including random effects/hierarchical modeling, OLS estimation would still yield unbiased and consistent coefficient estimates.[17] On the other hand, the composite residual's variance, $V(\varepsilon^*)$, is not constant (homoskedastic) but differs (heteroskedastic) across observations, even if $V(\varepsilon) \ldots V(\varepsilon_2)$ are each constant, rendering coefficient estimates and standard errors inefficient. Moreover, this nonconstant variance moves with the values of x and z, which implies that the standard-error estimates (but not the coefficient estimates) are biased and inconsistent as well. Thus, even if the error components in the random-effects model have constant variance,

---

17. $E(\hat{\boldsymbol{\beta}}_{ols}) = E((\mathbf{X}'\mathbf{X})^{-1}\mathbf{X}'\mathbf{y}) = E((\mathbf{X}'\mathbf{X})^{-1}\mathbf{X}'(\mathbf{X}\boldsymbol{\beta} + \varepsilon^*)) = \boldsymbol{\beta} + E((\mathbf{X}'\mathbf{X})^{-1}\mathbf{X}'\varepsilon^*) = \boldsymbol{\beta} + 0 = \boldsymbol{\beta}$ if each component of ε^* has mean zero and does not covary with $\mathbf{x}$. See Franzese (2005) for a fuller discussion of the proof.

mean zero, and no correlation with regressors, as we would commonly assume, OLS coefficient estimates will be inefficient, and OLS standard-error estimates will be biased, inconsistent, and inefficient. These problems, though potentially serious, are probably small in magnitude in most cases and, anyway, easy to redress by simple techniques with which political scientists are already familiar.

As mentioned before, similar issues arise in the literature on hierarchical, or multilevel, models (see, e.g., Bryk and Raudenbush 2001; Kedar and Shively 2005; Steenbergen and Jones 2002). Often these models propose that some unit-level y_{ij} depends on a contextual-level variable, z_j, varying only across and not within the j contexts, and a unit-level variable, x_{ij}, and furthermore that the effect of the unit-level variable x_{ij} depends (deterministically or stochastically) on z_j:

$$y_{ij} = \beta_0 + \beta_1 x_{ij} + \beta_2 z_j + \varepsilon_{ij} \tag{53}$$

$$\beta_0 = \gamma_0 (+\varepsilon_{0ij})$$

$$\beta_1 = \delta_1 + \delta_2 z_j (+\varepsilon_{1j})$$

$$\beta_2 = \delta_3 + \delta_4 x_{ij} (+\varepsilon_{2ij})$$

which implies that one may model y for regression analysis as

$$y = \gamma_0 + \beta_x x + \beta_z z + \beta_{xz} xz + \varepsilon^* \tag{54}$$

where $\varepsilon^* = \varepsilon_{ij} (+\varepsilon_{0ij} + \varepsilon_{1j} x_{ij} + \varepsilon_{2ij} z_j)$ and the coefficients remain identical to those given previously.

Assuming deterministic conditional relationships so that $\varepsilon^* = \varepsilon_{ij}$, that is, the parenthetical terms are all zero, and assuming that this simple residual is well behaved (mean zero, constant variance, and no correlation with regressors, as usual), OLS is BLUE. If, instead, ε_{ij} exhibits heteroskedasticity and/or correlation across i or j, then OLS coefficient and standard-error estimates would be unbiased and consistent but inefficient in the case that the patterns of these nonconstant variances and/or correlations were themselves uncorrelated with the regressors, their cross-products, and their squares. In the case that these patterns correlated in some fashion with the regressors, their cross-products, or their squares, OLS coefficient estimates would still be unbiased and consistent but inefficient, but OLS standard errors would be biased and inconsistent as well as inefficient in this context. These standard-error inconsistency problems could be redressed in a familiar manner by replacing the OLS formula for estimating the variance-covariance of estimated coefficients with a heteroskedasticity-consistent formula like White's or the appropriate heteroskedasticity-and-correlation-consistent formula, like Newey-West for temporal correlation, Beck-Katz for contemporaneous (spatial) correlation, or "cluster" for the case of common stochastic shocks to all units i in each context j.

With stochastic dependence such that $\varepsilon^* = \varepsilon_{ij} + \varepsilon_{0ij} + \varepsilon_{1j}x_{ij} + \varepsilon_{2ij}z_j$, on the other hand, OLS coefficient estimates are still unbiased and consistent, but the error term presents us with two issues even in the case of well-behaved ε_{ij}: heteroskedasticity (the composite residual term, ε^*, varies; in fact, it varies depending on some linear combination of x and z) as well as potentially severe autocorrelation (each ε_{1j} will be common to all individuals i in context j).[18]

Thus, the random-effects and multilevel (hierarchical) cases produce identical problems in OLS, and so the same solutions will apply. Note first that some form of the familiar White or Huber-White consistent variance-covariance estimators, that is, "robust" standard errors, will redress the inconsistency in OLS estimates of the estimated coefficients' variance-covariance, that is, $\widehat{V(\hat{\beta})}_{ols}$.

Recall that, given nonspherical disturbances,

$$V(\hat{\beta}) = E[(\beta - \hat{\beta})(\beta - \hat{\beta})'] = E[\{\beta - (\beta + (X'X)^{-1} X'\varepsilon)\}$$
$$\times \{\beta - (\beta + (X'X)^{-1}X'\varepsilon)\}']$$
$$= E[\{(X'X)^{-1}X'\varepsilon\}\{(X'X)^{-1} X'\varepsilon\}'] = E[(X'X)^{-1}X'\varepsilon\varepsilon'X(X'X)^{-1}]$$
$$= (X'X)^{-1}X' [E(\varepsilon\varepsilon')]X(X'X)^{-1} = (X'X)^{-1}X' [V(\varepsilon)]X(X'X)^{-1} \quad (55)$$

Under classical linear-regression assumptions, $\varepsilon \sim N(0, \sigma^2 I)$ and $E(\varepsilon'X) = 0$, and so this reduces to

---

18. Some current literature even suggests that OLS is biased in the presence of such multilevel random effects. This is false if *biased* refers to the OLS coefficient estimates. Provided that the context-specific or other components of the composite error term do not correlate with the regressors, OLS coefficient estimates will remain unbiased and consistent, although inefficient. The fact that Z_j and ε_j are both common to all individuals in context j implies that the pattern of the nonsphericity in the composite $V(\varepsilon^*)$ relates to a regressor, Z, producing biased, inconsistent, and inefficient OLS standard-error estimates, but that does not imply that $C(Z_j, \varepsilon^*)$ is nonzero, which is the condition that would bias OLS coefficient estimates. The "problem" with OLS for hierarchical models therefore resides solely in the inefficiency of OLS coefficient estimates and in the generally poor properties of the OLS estimates of $\widehat{V(\hat{\beta})}$. The problem is similar to that typically induced by strong temporal or spatial correlation: OLS coefficient estimates are unbiased and consistent but inefficient; standard errors are biased, inconsistent, and inefficient. The inefficiency in coefficient estimates can be dramatic if the within-context correlation of individual errors is great, perhaps dramatic enough to render unbiasedness and consistency of little practical comfort, but, even so, the problem is efficiency, not bias or inconsistency.

$$\mathbf{V}(\hat{\boldsymbol{\beta}}) = E[(\boldsymbol{\beta} - \hat{\boldsymbol{\beta}})(\boldsymbol{\beta} - \hat{\boldsymbol{\beta}})'] = (\mathbf{X}'\mathbf{X})^{-1}\mathbf{X}'[V(\boldsymbol{\varepsilon})](\mathbf{X}'\mathbf{X})^{-1}\mathbf{X}$$

$$= (\mathbf{X}'\mathbf{X})^{-1}\mathbf{X}'\sigma^2\mathbf{I}\mathbf{X}(\mathbf{X}'\mathbf{X})^{-1}$$

$$= \sigma^2(\mathbf{X}'\mathbf{X})^{-1}\mathbf{X}'\mathbf{X}(\mathbf{X}'\mathbf{X})^{-1} = \sigma^2(\mathbf{X}'\mathbf{X})^{-1}$$

With random effects, $\varepsilon^* = \varepsilon + \varepsilon_0 + \varepsilon_1 x + \varepsilon_2 z$; in multilevel data, $\varepsilon^* = \varepsilon_{ij} + \varepsilon_{0ij} + \varepsilon_{1j} x_{ij} + \varepsilon_{2ij} z_j$. Both violate the assumptions of classical linear regression in essentially the same way. In our random-coefficient case:

$$E(\boldsymbol{\varepsilon}^* \boldsymbol{\varepsilon}^{*'}) = E(\varepsilon + \varepsilon_0 + \varepsilon_1 x + \varepsilon_2 z)(\varepsilon + \varepsilon_0 + \varepsilon_1 x + \varepsilon_2 z)'$$

$$= E\begin{pmatrix} \varepsilon\varepsilon' + \varepsilon_0\varepsilon' + \varepsilon_1 x\varepsilon' + \varepsilon_2 z\varepsilon' + \varepsilon\varepsilon_0' + \varepsilon_0\varepsilon_0' + \varepsilon_1 x\varepsilon_0' \\ + \varepsilon_2 z\varepsilon_0' + \varepsilon x'\varepsilon_1' + \varepsilon_0 x'\varepsilon_1' + \varepsilon_1 xx'\varepsilon_1' + \varepsilon_2 zx'\varepsilon_1' \\ + \varepsilon z'\varepsilon_2' + \varepsilon_0 z'\varepsilon_2' + \varepsilon_1 xz'\varepsilon_2' + \varepsilon_2 zz'\varepsilon_2' \end{pmatrix} \quad (56)$$

Even assuming that $(\varepsilon, \varepsilon_0, \varepsilon_1, \varepsilon_2)$ are independently and identically distributed (i.i.d.) $N(0, \sigma^2\mathbf{I})$, the variance-covariance matrix for $\hat{\boldsymbol{\beta}}$ in the random coefficient model will be

$$\mathbf{V}(\hat{\boldsymbol{\beta}}_{RC}) = 2\sigma^2 + \mathbf{x}\mathbf{x}'\sigma^2 + \mathbf{z}\mathbf{z}'\sigma^2 = \sigma^2(2\mathbf{I} + \mathbf{x}\mathbf{x}' + \mathbf{z}\mathbf{z}') \quad (57)$$

In the hierarchical model, the basic structure is the same, but the claim that $(\varepsilon, \varepsilon_0, \varepsilon_1, \varepsilon_2)$ would be i.i.d. is less plausible because, among other reasons, context-level variance (ε_{1j}) is unlikely to equal unit-level variances $(\varepsilon_{ij}, \varepsilon_{0ij}, \varepsilon_{2ij})$. It is more plausible to assume that between-level variation differs but within-level variation is constant. If so, the variance-covariance of $\hat{\boldsymbol{\beta}}$ in the hierarchical case is

$$\mathbf{V}(\hat{\boldsymbol{\beta}}_{HM}) = 2\sigma^2_{ind} + \mathbf{x}\mathbf{x}'\sigma^2_{context} + \mathbf{z}\mathbf{z}'\sigma^2_{ind}$$

$$= \sigma^2_{ind}(2\mathbf{I} + \mathbf{z}\mathbf{z}') + \mathbf{x}\mathbf{x}'\sigma^2_{context} \quad (58)$$

Notice that the expressions for $\mathbf{V}(\hat{\boldsymbol{\beta}}_{HM})$ in the hierarchical case and for $\mathbf{V}(\hat{\boldsymbol{\beta}}_{RC})$ in the random-coefficient case are almost identical. The only difference is the separation we allow for the variances of components of ε^* in the hierarchical case, because such separation seems substantively sensible, that we do not allow in the random-coefficient case. In either case, the familiar class of robust estimators and/or reasonably familiar versions of feasible generalized least squares (FGLS) will redress OLS problems sufficiently in a relatively straightforward manner.

Recall that White's heteroskedastic-consistent estimator, for example, is

$$\widehat{\mathbf{V}(\hat{\boldsymbol{\beta}})} = n(\mathbf{X}'\mathbf{X})^{-1}\mathbf{S}_0(\mathbf{X}'\mathbf{X})^{-1} \quad \text{where } \mathbf{S}_0 = \frac{1}{n}\sum_{i=1}^{n} e_i^2 \mathbf{x}_i \mathbf{x}_i'$$

As Greene (2003) writes, White's estimator "implies that, without actually specifying the type of heteroskedasticity, we can still make appropriate inferences based on the results of least squares" (199). More precisely, White's estimator produces *consistent* estimates of the coefficient estimates' variance-covariance matrix in the presence of pure heteroskedasticity (nonconstant variance) whose pattern is somehow related to a pattern in $\mathbf{xx'}$, that is, to some pattern in the regressors, the regressors squared, or the cross-products of the regressors. Thus, in our pure random-coefficient case, White's estimator provides consistency ("robustness") to precisely the heteroskedasticity issue raised because the pattern of nonconstant variance depends on the regressors x and z and heteroskedasticity is the only issue raised. In the hierarchical-model case, we might additionally have concerns about a correlation among residuals due to the common components, ε_{1j}, in the errors of all individuals in context j. The pattern of this induced correlation will likewise relate to the regressors x and z (and their products and cross-products). In this case, a Huber-White heteroskedasticity-*and-clustering*-consistent variance-covariance estimator will produce the appropriately "robust" standard errors.[19]

Such "robust" standard-error estimators leave the inefficient coefficient estimates unchanged and are not efficient in their estimates of coefficient-estimate variance-covariance either. To redress these issues, feasible weighted least squares (FWLS) may be appropriate for the pure heteroskedasticity induced by simple random effects, and FGLS may be appropriate for the heteroskedasticity and correlation induced by the clustering likely in the hierarchical context. Specifically, since the patterns of heteroskedasticity or correlated errors producing the concerns are a simple function of the regressors involved in the interactions, one can conduct FWLS if appropriate and desired following these steps: (1) estimate by OLS; (2) save the OLS residuals; (3) square the OLS residuals; (4) regress the squared residuals on the offending regressors (x and z here); (5) save the predicted values of this auxiliary regression. The researchers would then (6) use the inverse of the square roots of these predicted values as weights for the FWLS reestimation. One may wish instead to regress the *log* of the squared OLS residuals on the offending regressors and save the *exponential* of these fitted values in step (5) to avoid estimating negative variances and then attempting to invert their square roots in step (6). The procedure for implementing FGLS if appropriate and desired is similar, except that both variance and covariance parameters are to be esti-

---

19. Again, Franzese (2005) discusses this matter further.

mated in steps (3) and (4) for insertion into the $\widehat{V(\hat{\varepsilon})}$ whose "square root inverse" is to provide the weighting matrix in step (6).[20]

As evidence in support of the claim that some form of a robust-cluster estimate will suffice in the hierarchical model with random coefficients case, we conducted several Monte Carlo experiments applying OLS, OLS with heteroskedasticity-consistent standard-error estimation, OLS with heteroskedasticity-and-cluster-consistent standard-error estimation, and random-effect-model estimation.[21] In all cases, the data were actually generated using hierarchical-model structures (with several alternative relative variances and covariances of the error components and the right-hand-side variables) and in samples with fifty *j* units and one hundred observations per unit (to correspond to a rather small survey conducted in each of the fifty U.S. states). All four estimation techniques yielded unbiased coefficient estimates, but the standard-error estimates, not surprisingly, were wrong with OLS and with robust standard-error estimates that ignore within-level autocorrelation (i.e., estimators consistent to heteroskedasticity only) but were nearly as good with the robust-cluster-estimation strategy as with the full random-effects model (the estimates were within 5 percent of each other). Appreciable efficiency gains in coefficient estimates from the hierarchical models relative to the OLS models were also notably absent. Accordingly, the main conclusion of our exercise was that one seemed generally to have little to gain—*in linear models in samples of these dimensions anyway*—from complicated random-coefficients and hierarchical-modeling strategies. OLS with robust variance-covariance estimator strategies (e.g., in STATA, one simply appends ", robust" or ", robust cluster" to the end of the estimation command) seemed generally to suffice. Of course, we would demand much further simulation, across wider and more systematically varying model types and ranges of parameters and sample dimensions, to support this conclusion more wholeheartedly as a general one. In this sample dimension and model context at least, however, simpler strategies work almost indistinguishably from the more

---

20. The "square-root inverse" of a matrix with nonzero off-diagonal elements is not a simple inversion of the square root of each of the elements, as it is in the FWLS case where $V(\varepsilon)$ is diagonal. However, most statistical software packages will find the square-root inverse of a matrix, and so we need not detain the reader with these computations.

One could also iterate the FWLS or FGLS procedures, and common practice is to do so, even though, statistically, the iterated and one-shot strategies have identical properties.

21. The variance-covariance matrix for coefficients estimated with the particular robust cluster we implemented (using STATA) is $\widehat{V(\hat{\beta})} = (X'X)^{-1}S_J(X'X)^{-1}$ where $S_J = \sum_{j=1}^{J}\mathbf{u}_j'\mathbf{u}_j$ and where $\mathbf{u}_j = \sum_{i=1}^{n_j}e_{ij}\mathbf{x}_{ij}$. We estimated the random effects model using hierarchical linear model (HLM) software.

complex ones, and so we are happy to argue for simplicity in cases like this at any rate. We also note, however, that the properties of these "robust" standard-error estimators deteriorate in smaller samples. For the simple heteroskedasticity-consistent estimator, this seems to occur only in very small samples beginning around $N = 35$. For robust-cluster estimators, two sample-size dimensions are key: total, N, and J, the number of "contexts." Again, very small J, say, below about thirty, and/or N become increasingly problematic.[22]

---

22. These sample sizes and dimensions come from consideration of the small-sample adjustments some statisticians have recommended to these robust estimators, multiplying White's by a term involving $N/(N - 1)$ and robust cluster by a term involving $[N/(N - 1)][J/(J - 1)]$. Franzese (2005) discusses these considerations in far greater depth. See also Achen (2005), who correctly stresses the possible reliance upon linearity for many of these results and conclusions.

6

≋

SUMMARY

We have emphasized the importance of understanding the links between substantive theories and empirical tests of those theories.
Social scientists often formulate hypotheses that demand some complexity beyond the simple linear-additive model. Multiplicative interaction
terms provide one simple means to enhance the match of these complex
theories to appropriate empirical statistical analyses.

We conclude with this summary of our recommendations on the use
and interpretation of interactive terms in linear-regression models. In
order:

- *Theory:* What is the scientific phenomenon to be studied? Does your
  theory suggest that the effects of some variable(s) **x** depend on some
  other variable(s) **z** (implying the converse that the effect(s) of **z** de-
  pend(s) on **x**)? Does it imply anything more specific about the manner
  in which the effects of **x** and of **z** depend on each other?
- *Model:* What is the appropriate mathematical model to express your
  theory? Write the formal mathematical expression that encapsulates
  your theory. In the case where the theory implies that the effect(s) of **x**
  depend(s) on **z** and vice versa, (a) simple multiplicative interaction(s)
  will often suffice to express that (those) proposition(s). If the theory
  implies something more specific, ideally one would specify that more
  specific (perhaps nonlinear) form of the interactions.
- *Estimation:* Estimate the model with an appropriate estimation strategy; OLS (or nonlinear regression model) with appropriately "robust"
  standard errors typically suffices.

131

- *Interpretation:* What are the substantive effects of interest? Conduct appropriate hypothesis tests that match your substantive theoretical propositions. Calculate marginal effects using derivatives to describe the effects of the variable(s) of interest, x and/or z, at various, meaningful levels of the other variables. Calculate changes in the predicted values of y induced as some variable(s) of interest, x and/or z, change(s) at various, meaningful levels of the other variables. Also calculate the standard errors of these estimated effects and/or confidence intervals.
- *Presentation:* Present tables or graphs including both marginal effects or differences and accompanying measures of uncertainty or including both predicted values and accompanying measures of uncertainty. Plot or tabulate these effects across a range of meaningful levels of the other variables.

APPENDIX A.
DIFFERENTIATION RULES

Here is a table of useful differentiation rules (for a more complete list of differentiation rules, we refer the reader to Kleppner and Ramsey 1985).

Let a, b, c = constants; x, z, w = variables; y = a function of some variable(s); $f(\ )$, $g(\ )$ = functions.

TABLE A1. Some Useful Differentiation Rules

Expression	$\partial y/\partial x$	Explanation	Example
$y = c$	$\partial c/\partial x = 0$	The derivative of a constant is zero.	$\partial 7/\partial x = 0$
$y = cz$	$\partial(cz)/\partial x = 0$	The derivative of a term that does not depend on x is zero.	$\partial(3z)/\partial x = 0$
$y = cx$	$\partial(cx)/\partial x = c$	The derivative of a term involving a linear coefficient and x is that coefficient.	$\partial(3x)/\partial x = 3$
$y = cx^a$	$\partial(cx^a)/\partial x = acx^{a-1}$	The derivative of a term involving a linear coefficient and x raised to the ath power is the product of a, c, and x raised to the $(a-1)$ power.	$\partial(3x^5)/\partial x = 15x^4$
$y = cxz$	$\partial(cxz)/\partial x = cz$	The derivative of a term involving a linear coefficient, x, and another variable, z, is the product of the coefficient and the variable (we can treat the other variable as a constant with respect to x here).	$\partial(3xz)/\partial x = 3z$
$y = cxzw$	$\partial(cxzw)/\partial x = czw$	The result extends to higher order interactions, where again variables that are not a function of the variable with respect to which one is differentiating are fixed.	$\partial(3xzw)/\partial x = 3zw$

$y = \ln(x)$ $\partial(\ln(x))/\partial x = 1/x$	The derivative of a logged variable is the inverse of that variable.	$\partial(3\ln(x))/\partial x = 3/x$
$y = e^x$ $\partial(e^x)/\partial x = e^x$	The derivative of base e raised to a variable is base e raised to that variable.	$\partial(3e^x)/\partial x = 3e^x$
$y = b_0 + b_x x + b_z z + b_{xz} xz$ $\partial b_0/\partial x + \partial(b_x x)/\partial x + \partial(b_z z)/\partial x + \partial(b_{xz} xz)/\partial x = b_x + b_{xz} z$	The derivative of some linear-additive function equals the sum of the derivative of each of the terms.	$\partial(1 + 2x + 3z + 4xz)/\partial x = 2 + 4z$
$y = f(x) \times g(x)$ $\partial(f(x) \times g(x))/\partial x = \partial(f(x))/\partial x\, g(x) + \partial(g(x))/\partial x\, f(x)$	The derivative of the product of two functions equals the sum of derivative of the first function, multiplied by the undifferentiated second function; plus the derivative of the second function, multiplied by the undifferentiated first function.	$\partial((2x + 5) \times (3\ln(x)))/\partial x = \partial((2x + 5))/\partial x\,(3\ln(x)) + \partial(3\ln(x))/\partial x\,(2x + 5) = 2(3\ln(x)) + (3/x)(2x + 5)$
$y = f(g(x))$ $(df/dg) \times (dg/dx)$	This is the chain rule for nested functions.	$\partial((2(3\ln x) + 5))/\partial x = \partial((2(g) + 5))/\partial g \times \partial g/\partial x = 2 \times (3/x) = 6/x$
$F \equiv$ a cumulative probability function for the probability density function f. $\partial F(x)/\partial x = f(x)$	The derivative of any cumulative probability function is the corresponding probability density function.	$\partial\Phi(x)/\partial x = \phi(x)$

APPENDIX B.
STATA SYNTAX

M any statistical software packages are available to researchers. Because STATA is prominent in the social sciences, we provide STATA-
based syntax for readers to use in following our advice on interpreting and
presenting results from linear models that include interaction terms.[1]

We advise creating a separate data set that contains simulated values
for each of the variables in the regression analysis (it can be used for marginal effects and/or for predicted values, or separate ones can be used for
each approach). A data set of simulated values enables the researcher to
interpret effects along evenly spaced values of one or more of the variables, within a substantively useful range at which marginal effects, predicted values, and differences in predicted values can easily be interpreted (where the actual data set may not contain evenly spaced values).

Marginal Effects, Standard Errors,
and Confidence Intervals

To begin, determine the number of observations that will be contained in
the simulation data set. A researcher might want to calculate the estimated
marginal effects of x as z ranges from its minimum to its maximum, at

---

1. This syntax is valid for STATA version 9.

evenly spaced increments (e.g., if the variable z ranges from 1 to 10, and the user wishes z to vary in 1-unit increments, this would imply 10 observations). We advise selecting a manageable number of values (5–100). Open STATA and create a new data set by setting the number of observations, v (e.g., "10"), to be included:

```
set obs v
```

One could manually enter each evenly spaced value into a data set (e.g., 1, 2, 3, etc.) using the data editor. A more efficient way of setting values of z is easily found:

```
egen z = fill(min(unit)max)
```

This command creates a variable z that ranges from *min* (e.g., "1") to *max* (e.g., "10") in *unit* increments (e.g., "1"). If, following our government-durability example, z is to range between 40 and 80 and rise by 5-unit increments, then we would need 9 observations, and we would run the following command lines to generate values of z:

```
set obs 9

egen z = fill(40(5)80)
```

Then, save the data set:

```
save dydxdata.dta
```

After you have created a data set that allows for a range of z values, return to the empirical data:

```
use realdata.dta
```

To estimate the following "standard model," given variables y, x, z and other covariates, w, in the data set:

$$y = \beta_0 + \beta_x x + \beta_z z + \beta_{xz} xz + \beta_w w + \varepsilon$$

Generate the multiplicative term, xz:

```
gen xz = x*z
```

Estimate the linear-regression model:

```
regress y x z xz w
```

Recall that the marginal effects of x and z on y are $\partial \hat{y}/\partial x = \hat{\beta}_x + \hat{\beta}_{xz} z$ and $\partial \hat{y}/\partial z = \hat{\beta}_z + \hat{\beta}_{xz} x$.

Marginal effects are calculated by adding the estimated $\hat{\beta}_x$ to the product of each value of z with the estimated coefficient $\hat{\beta}_{xz}$. Open the simulation data set to calculate marginal effects:

use dydxdata.dta, clear

This command will call up the simulation data set and clear the empirical data set. The OLS estimates will remain in memory (although typing "clear" on its own *will* remove the estimates from memory).

One could take the estimated coefficients from the regression output and create a new variable:

gen dydx = $\hat{\beta}_x$ + z* $\hat{\beta}_{xz}$

where the estimated value $\hat{\beta}_x$ from the regression output (e.g., "-2") is entered instead of "$\hat{\beta}_x$" and the estimated value of $\hat{\beta}_{xz}$ from the regression output (e.g., "10") is entered in place of "$\hat{\beta}_{xz}$." This command line would thus create v values corresponding with the marginal effects of x at various values of z. A disadvantage to this procedure is that it is possible to mistype the estimated values, creating grave errors in the calculated effects. A less error-prone way of calculating the marginal effects, then, is to use the estimates STATA stores in memory.

STATA stores the estimated coefficient $\hat{\beta}_x$ in memory as _b[x] and the coefficient $\hat{\beta}_{xz}$ in memory as _b[xz], and so a variable that consists of the marginal effects of x as z varies across the evenly incrementing values of z is generated as follows:

gen dydx=_b[x]+_b[xz]*z

Using the variable dydx, a table or plot of selected marginal effects for evenly spaced values of interest is now easily created.[2]

Presentations of marginal effects should also include an indication of our level of certainty or uncertainty regarding these marginal effects. Recall that the estimated variance of the marginal effects in this example would be $\widehat{V(\partial\hat{y}/\partial x)} = \widehat{V(\hat{\beta}_x)} + z^2\widehat{V(\hat{\beta}_{xz})} + 2z\widehat{C(\hat{\beta}_x,\hat{\beta}_{xz})}$. Calculating $\widehat{V(\partial\hat{y}/\partial x)}$ is straightforward from there. The estimated variance-covari-

---

2. Users can also take advantage of STATA's programmed postestimation commands. The command lincom will report estimates, standard errors, t-statistics, p-levels, and a 95 percent confidence interval for any linear combination of coefficients. So, lincom can be used to calculated marginal effects at selected values of z: lincom x zvalue*xz will calculate $\hat{\beta}_x + \hat{\beta}_{xz}z$ at the z-value entered into the command line. For a handful of marginal effects, lincom is a useful shortcut; the disadvantage is that the z-values must be entered one at a time. If more than a handful of effects are desired (or if graphing is desired), then the procedure outlined in the text will be more serviceable. Alternatively, lincom can be written into a looping program and the results stored in a data set that will allow graphing. Appendix B contains this programming syntax.

ance matrix of estimated coefficients is retrieved in STATA by typing "vce" after an estimation command. The user could then simply generate a new variable by taking the specific values of $\widehat{V(\hat{\beta}_x)}$, $\widehat{V(\hat{\beta}_{xz})}$, and $\widehat{C(\hat{\beta}_x,\hat{\beta}_{xz})}$ acquired from viewing the values in the variance-covariance matrix.

gen vardydx = $\widehat{V(\hat{\beta}_x)}$+z*z*$\widehat{V(\hat{\beta}_{xz})}$+2*z*$\widehat{C(\hat{\beta}_x,\hat{\beta}_{xz})}$

where $\widehat{V(\hat{\beta}_x)}$, $\widehat{V(\hat{\beta}_{xz})}$, and $\widehat{C(\hat{\beta}_x,\hat{\beta}_{xz})}$, would be replaced by their estimated values (e.g., "2").

Again, although this "enter by hand" method is transparent, human error in data entry could be a problem. A less error-prone method uses the estimates that STATA stores in memory. The square root of $\widehat{V(\hat{\beta}_x)}$ is in "_se[x]," and the square root of $\widehat{V(\hat{\beta}_{xz})}$ is in "_se[xz]." The value $\widehat{C(\hat{\beta}_x,\hat{\beta}_{xz})}$ is stored as the element in the row and column corresponding to x and xz in the estimated variance-covariance matrix of the coefficient estimates, vce. In this particular case, given the order of the variables in the estimated equation, it is in the third row, first column (and, because the variance-covariance matrix is symmetric, also in the first row, third column).

Create a matrix V to represent the variance-covariance matrix of the coefficient estimates, VCE.

matrix V = get(VCE)

Generate a variable C_x_xz, which contains the covariance of interest, extracted from its position in the matrix V.

gen C_x_xz = V[3,1]

The estimated variance (and standard error) of each estimated marginal effect can thus be calculated as

gen vardydx=(_se[x]^2)+(z*z)*(_se[xz]^2)+2*z*C_x_xz

gen sedydx=sqrt(vardydx)

A table of marginal effects with accompanying standard errors could be generated as follows:

tabdisp z, cellvar(dydx sedydx)

This command line would present a table featuring all v values of z, with the appropriate marginal effect and standard error of the marginal effect. This table is likely to be useful for the researcher for interpretation, but for presentational purposes, only a set of selected values of z might be incorporated into an abbreviated table.

Alternatively, marginal effects can be graphed. Recall that confidence intervals can be generated with the following formula: $\partial \hat{y}/\partial x \pm t_{df,p}$ $\sqrt{\widehat{V(\partial \hat{y}/\partial x)}}$. STATA stores the degrees of freedom from the previous estimation as "e(df_r)," and the researcher can use the inverse t-distribution function to create $t_{df,p}$. For a 95 percent confidence interval, the lower and upper bounds are calculated as

gen LBdydx=dydx-invttail(e(df_r),.025)*sedydx

gen UBdydx=dydx+invttail(e(df_r),.025)*sedydx

This command graphs estimated marginal effects with confidence intervals, along values of z:

twoway connected dydx LBdydx UBdydx z

These procedures are summarized in table B1.

TABLE B1.   STATA Commands for Calculating Marginal Effects of x on y, Standard Errors of Those Effects, and Confidence Intervals around Those Effects

Procedures	Command Syntax
Create simulation data set: v evenly spaced values for z from its minimum to its maximum. Save the data set.	set obs v egen z = fill(*min(unit)max*) save dydxdata.dta
Open the original data, generate the multiplicative term, and estimate the linear-regression model.	use realdata.dta gen xz = x*z regress y x z xz w
Open the simulation data set and calculate the estimated marginal effect.	use dydxdata.dta, clear gen dydx =_b [x] + _b [xz] *z
Create a matrix from the estimated covariance matrix of the coefficient estimates, pull out the stored element $C(\hat{\beta}_x, \hat{\beta}_{xz})$, and create a variable containing it.	matrix V = get (VCE) gen C_x_xz=V [3,1]
Calculate the estimated variance (and standard error) of each estimated marginal effect.	gen vardydx = (_se[x] ^2)+ (z*z) * (_se[xz] ^2) +2*z*C_x_xz gen sedydx = sqrt (vardydx)
Generate a table displaying estimated marginal effects and standard errors for all v values of z.	tabdisp z, cellvar(dydx sedydx)
Generate lower and upper confidence-interval bounds.	gen LBdydx = dydx-invttail (e(df_r),.025)*sedydx gen UBdydx = dydx + invttail(e(df_r),.025)*sedydx
Graph the estimated marginal effects and the upper and lower confidence intervals along values of z.	twoway connected dydx LBdydx UBdydx z

Differences in predicted values can be generated by multiplying the marginal effects calculated previously by Δx (recall that $\Delta y = \Delta x(\hat{\beta}_x + \hat{\beta}_{xz}z_0)$ so long as x enters linearly into the interaction). The estimated variance of these differences in predicted values, similarly, is calculated by multiplying the estimated variance of the estimated marginal effect by Δx^2. For example:

gen diffyhat = a*(dydx)

gen vardiffyhat = (a^2)*vardydx

where a is Δx, (e.g., "2").

Predicted Values, Standard Errors, and Confidence Intervals

Predicted values, $\hat{y}$, are generated by summing the products of the right-hand-side variables, set at particular values, and their corresponding co-efficients: $\hat{y} = M_h\hat{\beta}$, where M_h is a matrix of values at which x, z, and any other variables in the model are set.

We advise creating a simulation data set that contains the values at which x, z, and any other variables in the model are to be set. Begin by determining the number of observations that will be contained in the data set. A researcher might want to calculate the estimated predicted values as z ranges from its minimum to its maximum, at evenly spaced increments (e.g., if the variable z ranges from 1 to 10, and the user wishes z to vary in 1-unit increments, this would imply 10 observations). We advise selecting a manageable number of values (5–100). Open STATA and create a new data set by setting the number of observations, v, to be included:

set obs v

One could manually enter each evenly spaced value into a data set (e.g., 1, 2, 3, etc.) using the data editor. A more efficient way of setting values of z is easily found:

egen z = fill(*min(unit)max*)

This command creates a variable z that ranges from *min* (e.g., "1") to *max* (e.g., "10") in *unit* increments (e.g., "1"). Set the other variables at the desired level, for example, the means, or modes, or substantively interesting values, using the generate command. To generate predicted probabilities based on a model that contains k regressors (including the constant), k total variables must be created. In this example, we create variable x that takes the value $c1$ (e.g., "10"), variable w that takes the value $c2$ (e.g., "-2"), and variable col1 that takes the value of 1 (later to be multiplied by the intercept).

gen x=$c1$

gen w=$c2$

gen col1=1

Note that each of these variables in the data set will be set at a constant value: the only variable that will vary is z; all other variables (aside from the interaction term) are held constant.[3] Create the interaction term that reflects the values to which x and z are held and save the data set.

gen xz=x*z

save yhatdata.dta

Open the real data set:

use realdata.dta

To estimate the following "standard model," given variables y, x, and z in the data set.

$$y = \beta_0 + \beta_x x + \beta_z z + \beta_{xz} xz + \varepsilon$$

Generate the multiplicative term, xz:

gen xz = x*z

Estimate the linear-regression model:

regress y x z xz w

Open the simulation data set:

use yhatdata.dta, clear

This command will call up the simulation data set and clear the empirical data set. The OLS estimates will remain in memory (although typing "clear" on its own *will* remove the estimates from memory).

---

3. To generate several blocks of set values that allow z to range from its minimum to its maximum, but also allow x to take on different values, the user could take advantage of the expand command. Generate the first block of values following the preceding instructions and setting x to x_a. Then expand the data set by two: expand 2. This command line will create a block of v additional observations that will exactly match the first. Then replace the value of x in the new block of observations: replace x=*newvalue* in $(v+1)/2v$ (e.g., replace x = 4 in 11/20).

Assemble the variables into a matrix:[4]

mkmat x z xz w col1, matrix(Mh)

This command creates matrix Mh, which contains the specified values at which our variables are set: x is fixed at the value $c1$; z varies at regular intervals between some minimum and maximum; the xz correspondingly varies, as it is the product of x (which is held at $c1$) and z (which varies). The covariate w is fixed at $c2$.

Recall that $\hat{y} = M_h\hat{\beta}$. Although $\hat{\beta}$ is a column vector of coefficients, STATA stores the estimated coefficients as a $1 \times k$ row vector, e(b). So we want to create B, a column vector with $k \times 1$ dimensions, that takes the stored coefficients and transposes them into $\hat{\beta}$:

matrix B=e(b)'

Calculating the predicted values is simply a matter of multiplying Mh by B:

matrix yhat=Mh*B

Then convert the resulting column vector into a variable, yhat:

svmat yhat, name(yhat)

Recall that $\widehat{V(\hat{y})} = M_h \widehat{V[\hat{\beta}]}M_h'$. STATA stores the estimated variance-covariance matrix of the estimated coefficients, $\widehat{V[\hat{\beta}]}$, as VCE in its memory. We create a matrix V consisting of $\widehat{V[\hat{\beta}]}$:

matrix V=get(VCE)

We can now calculate the variance of the predicted values as follows:

matrix VCEYH=Mh*V*Mh'

This command creates a matrix, VCEYH, that contains the variances and covariances of the predicted values. The diagonal elements in the variance-covariance matrix of predicted values are those of interest to us, as they correspond with the estimated variance of the predicted values. We want to extract these diagonal elements into a vector:

---

4. A shortcut is provided by the predict command. Estimate the regression on the original data, open the simulated data, and enter predict yhat (bypassing creation of **Mh**). The predict command line generates predicted values using the stored regression coefficients and the values of the variables in the current data set. As long as the variables in the simulation data set have the same name as the variables in the original data set, the predict command line will produce the desired results. Entering predict seyhat, stdp will generate standard errors around the predicted values.

```
matrix VYH= (vecdiag(VCEYH))'
```

The vecdiag command creates a row vector from the diagonal elements of the variance-covariance matrix of the predicted values. Because we want a column vector rather than a row vector, a transpose appears at the end of the command.

We then create a new variable, vyhat, which contains a unique estimated variance to correspond with each predicted value yhat:

```
svmat VYH, name(vyhat)
```

Taking the square root produces the estimated standard error of each predicted value yhat:

```
gen seyhat = sqrt(vyhat)
```

The researcher can next present a table of predicted values with corresponding standard errors:

```
tabdisp z, cellvar(yhat seyhat)
```

Predicted values are effectively displayed when graphed with confidence intervals. The confidence intervals around predicted values $\hat{y}$ can be constructed as $\hat{y} \pm t_{df,p} \sqrt{\widehat{V(\hat{y})}}$, where $\hat{y}$ corresponds with the values in yhat, $\sqrt{\widehat{V(\hat{y})}}$ corresponds with the values in seyhat, and $t_{df,p}$ refers to the relevant critical value from the t-distribution. STATA stores the degrees of freedom from the previous estimation as "e(df_r)," and the researcher can utilize the inverse t-distribution function to create the multiplier $t_{df,p}$.

For a 95 percent confidence interval, the lower and upper bounds are calculated as follows:

```
gen LByhat=yhat-invttail(e(df_r),.025)*seyhat
```

```
gen UByhat=yhat+invttail(e df_r),.025)*seyhat
```

The predicted values and confidence intervals can be graphed along values of z as follows:

```
twoway connected yhat LByhat UByhat z
```

These procedures are summarized in table B2.

TABLE B2. STATA Commands for Calculating Predicted Values of y, Standard Errors for Those Predicted Values, and Confidence Intervals around Those Predicted Values

Procedures	Command Syntax
Create simulation data set that contains k total variables. Begin with v evenly spaced values for z from its minimum to its maximum. Create variables that set the remaining covariates at meaningful values, including a column of 1s for the intercept. Create the interaction term. Save the data set.	set obs v egen z = fill($min(unit)max$) gen x = $c1$ gen w = $c2$ gen col1 = 1 gen xz = xz save yhatdata.dta
Open the original data, generate the multiplicative term, and estimate the linear-regression model.	use realdata.dta gen xz = x*z regress y x z xz w
Open the simulation data set and assemble the matrix of set values.	use yhatdata.dta, clear mkmat x z xz w col1, matrix(Mh)
Create a column vector containing the coefficient estimates.	matrix B = e(b)'
Create a column vector of predicted values.	matrix Yhat = Mh*B
Convert the column vector into a variable.	svmat Yhat, name(yhat)
Create a matrix from the estimated covariance matrix of the coefficient estimates.	matrix V = get(VCE)
Calculate the variance of the predicted values.	matrix VCEYH = Mh*V*Mh'
Extract the diagonal elements of the variance-covariance matrix of predicted values into a column vector.	matrix VYH = (vecdiag(VCEYH))'
Convert the column vector into a variable.	svmat VYH, name(vyhat)
Calculate the estimated standard error of each predicted value.	gen seyhat = sqrt(vyhat)
Present a table of predicted values with corresponding standard errors.	tabdisp z, cellvar(yhat seyhat)
Generate lower and upper confidence-interval bounds.	gen LByhat = yhat-invttail(e(df_r),.025)*seyhat gen UByhat = yhat + invttail(e(df_r),.025)*seyhat
Graph the predicted values and the upper and lower confidence intervals along values of z.	twoway connected yhat LByhat UByhat z

Marginal Effects, Using "lincom"

The STATA command "lincom" provides a shortcut for calculating marginal effects and their estimated standard errors. It calculates a linear combination of estimators following regression. The disadvantage to "lincom" is that it can be cumbersome to use when the user desires to calculate marginal effects across several values. Here, we provide a looping command that applies lincom across a range of values. In the example, the marginal effects of x are calculated across values of z (for clarity, denoted as zvalues), as z takes values between 0 and 6. The programming loop will post four types of results to a data set called lincomresults.dta: the marginal effect estimates from lincom, the associated standard errors, the value of zvalues applied, and the degrees of freedom in the model (this will be constant throughout, but it helps to have STATA collect it).

```
program define lincomrange
        version 9
        tempname dydx
        postfile 'dydx' dydx sedydx zvalues df using lincomresults, replace
        quietly {
                forvalues z = 0/6 {
                        drop _all
                        use realdata.dta
                        reg y x z xz w
                        lincom x + 'zvalues'*xz
                        post 'dydx' (r(estimate)) (r(se)) ('zvalues') (e(df_r))
                        }
                }
postclose 'dydx'
end
lincomrange
use lincomresults, clear
tabdisp zvalues, c(dydx sedydx)
gen LBdydx = dydx-invttail (df,.025)*sedydx
gen UBdydx = dydx+invtttail (df,.025)*sedydx
twoway connected dydx LBdydx UBdydx zvalues
```

REFERENCES

Achen, Christopher H. *Interpreting and Using Regression*. Thousand Oaks, CA: Sage, 1982.

Achen, Christopher H. "Two-Step Hierarchical Estimation: Beyond Regression Analysis." *Political Analysis* 13, no. 4 (2005): 447–56.

Allison, Paul D. "Testing for Interaction in Multiple Regression." *American Journal of Sociology* 83, no. 1 (1977): 144–53.

Althauser, Robert P. "Multicollinearity and Non-Additive Regression Models." In *Causal Models in the Social Sciences*, ed. H. M. Blalock, Jr., 453–72. Chicago: Aldine Atherton, 1971.

Amorim Neto, Octavio, and Gary W. Cox. "Electoral Institutions, Cleavage Structures, and the Number of Parties." *American Journal of Political Science* 41, no. 1 (1997): 149–74.

Baron, Reuben M., and David A. Kenny. "The Moderator-Mediator Variable Distinction in Social Psychological Research: Conceptual, Strategic, and Statistical Considerations." *Journal of Personality and Social Psychology* 51, no. 6 (1986): 1173–82.

Beck, Nathaniel, Kristian Gleditsch, and Kyle Beardsley. "Space Is More than Geography: Using Spatial Econometrics in the Study of Political Economy." *International Studies Quarterly* 50, no. 1 (2006): 27–44.

Berry, Frances Stokes, and William D. Berry. "State Lottery Adoptions as Policy Innovations: An Event History Analysis." *American Political Science Review* 84, no. 2 (1990): 395–415.

Berry, Frances Stokes, and William D. Berry. "Specifying a Model of State Policy Innovation." *American Political Science Review* 85, no. 2 (1991): 573–79.

Box-Steffensmeier, Janet M., Suzanna De Boef, and Tse-min Lin. "The Dynamics of the Partisan Gender Gap." *American Political Science Review* 98, no. 3 (2004): 515–28.

Bryk, Anthony S., and Stephen W. Raudenbush. *Hierarchical Linear Models: Applications and Data Analysis Methods*. 2nd ed. Newbury Park, CA: Sage, 2001.

147

Burns, Nancy. "Gender: Public Opinion and Political Action." In *Political Science: The State of the Discipline,* ed. Ira Katzelson and Helen V. Milner, 462–87. New York: Norton, 2002.

Cox, Gary W. *Making Votes Count: Strategic Coordination in the World's Electoral Systems.* Cambridge, UK: Cambridge University Press, 1997.

Cronbach, Lee J. "Statistical Tests for Moderator Variables: Flaws in Analyses Recently Proposed." *Psychological Bulletin* 102, no. 3 (1987): 414–17.

Dunlap, William P., and Edward R. Kemery. "Failure to Detect Moderating Effects: Is Multicollinearity the Problem?" *Psychological Bulletin* 102, no. 3 (1987): 418–20.

Fisher, Gene A. "Problems in the Use and Interpretation of Product Variables." In *Common Problems/Proper Solutions: Avoiding Error in Quantitative Research,* ed. J. Scott Long, 84–107. Newbury Park, CA: Sage, 1988.

Frant, Howard. "Specifying a Model of State Policy Innovation." *American Political Science Review* 85, no. 2 (1991): 571–73.

Franzese, Robert J., Jr. "Partially Independent Central Banks, Politically Responsive Governments, and Inflation." *American Journal of Political Science* 43, no. 3 (1999): 681–706.

Franzese, Robert J., Jr. *Macroeconomic Policies of Developed Democracies.* Cambridge, UK: Cambridge University Press, 2002.

Franzese, Robert J., Jr. "Multiple Hands on the Wheel: Empirically Modeling Partial Delegation and Shared Control of Monetary Policy in the Open and Institutionalized Economy." *Political Analysis* 11, no. 4 (2003a): 445–74.

Franzese, Robert J., Jr. "Strategic Interactions of the ECB, Wage/Price Bargainers, and Governments: A Review of Theory, Evidence, and Recent Experience." In *Institutional Conflicts and Complementarities: Monetary Policy and Wage Bargaining Institutions in EMU,* ed. Robert J. Franzese, Jr., Peter Mooslechner, and Martin Schürz, 5–42. New York: Kluwer, 2003b.

Franzese, Robert J., Jr. "Empirical Strategies for Various Manifestations of Multilevel Data." *Political Analysis* 13, no. 4 (2005): 430–46.

Franzese, Robert J., Jr., and Jude Hays. Spatial Econometric Models for Political Science. Working paper, University of Michigan, Ann Arbor, and University of Illinois, Urbana-Champaign, 2005.

Friedrich, Robert J. "In Defense of Multiplicative Terms in Multiple Regression Equations." *American Journal of Political Science* 26, no. 4 (1982): 797–833.

Greene, William H. *Econometric Analysis.* 5th ed. Upper Saddle River, NJ: Prentice-Hall, 2003.

Hall, Peter A. *Governing the Economy: The Politics of State Intervention in Britain and France.* Cambridge, UK: Polity, 1986.

Hayduk, Leslie A., and Thomas H. Wonnacott. "'Effect Equations' or 'Effect Coefficients': A Note on the Visual and Verbal Presentation of Multiple Regression Interactions." *Canadian Journal of Sociology* 5, no. 4 (1980): 399–404.

Ikenberry, G. John. "Conclusion: An Institutional Approach to American Foreign Economic Policy." *International Organization* 42, no. 1 (1988): 219–43.

Jaccard, James, Robert Turrisi, and Choi K. Wan. *Interaction Effects in Multiple Regression.* Newbury Park, CA: Sage, 1990.

Jusko, Karen Long, and W. Phillips Shively. "Applying a Two-Step Strategy to the

Analysis of Cross-National Public Opinion Data." *Political Analysis* 13, no. 4 (2005): 327–44.

Kam, Cindy D., Robert J. Franzese, Jr., and Amaney A. Jamal. "Modeling Interactive Hypotheses and Interpreting Statistical Evidence Regarding Them." Paper presented at the 1999 Annual Meeting of the American Political Science Association, Atlanta, 1999.

Kedar, Orit, and W. Phillips Shively, eds. *Political Analysis: Special Issue on Multilevel Modeling for Large Clusters.* 13 (2005).

King, Gary, Michael Tomz, and Jason Wittenberg. "Making the Most of Statistical Analyses: Improving Interpretation and Presentation." *American Journal of Political Science* 44, no. 2 (2000): 347–62.

Kleppner, Daniel, and Norman Ramsey. *Quick Calculus.* 2nd ed. New York: John Wiley and Sons, 1985.

Lijphart, Arend. *Electoral Systems and Party Systems: A Study of Twenty-Seven Democracies, 1945–1990.* Oxford: Oxford University Press, 1994.

Lohmann, Susanne. "Optimal Commitment in Monetary Policy: Credibility versus Flexibility." *American Economic Review* 82, no. 1 (1992): 273–86.

Morris, J. H., J. D. Sherman, and E. R. Mansfield. "Failures to Detect Moderating Effects with Ordinary Least Squares-Moderated Multiple Regression: Some Reasons and a Remedy." *Psychological Bulletin* 99 (1986): 282–88.

Nagler, Jonathan. "The Effect of Registration Laws and Education on U.S. Voter Turnout." *American Political Science Review* 85, no. 4 (1991): 1393–405.

Ordeshook, Peter, and Olga Shvetsova. "Ethnic Heterogeneity, District Magnitude, and the Number of Parties." *American Journal of Political Science* 38, no. 1 (1994): 100–123.

Shapiro, Robert Y., and Harpreet Mahajan. "Gender Differences in Policy Preferences: A Summary of Trends from the 1960s to the 1980s." *Public Opinion Quarterly* 50, no. 1 (1986): 42–61.

Shepsle, Kenneth. "Studying Institutions: Some Lessons from the Rational Choice Approach." *Journal of Theoretical Politics* 1 (1989): 131–47.

Smith, Kent W., and S. W. Sasaki. "Decreasing Multicollinearity: A Method of Models with Multiplicative Functions." *Sociological Methods and Research* 8 (1979): 35–56.

Southwood, Kenneth E. "Substantive Theory and Statistical Interaction: Five Models." *American Journal of Sociology* 83, no. 5 (1978): 1154–203.

Steenbergen, Marco R., and Bradford S. Jones. "Modeling Multilevel Data Structures." *American Journal of Political Science* 46, no. 1 (2002): 218–37.

Steinmo, Sven, Kathleen Thelen, and Frank Longstreth, eds. *Historical Institutionalism in Comparative Politics.* Cambridge, UK: Cambridge University Press, 1992.

Stolzenberg, Ross M. "The Measurement and Decomposition of Causal Effects in Nonlinear and Nonadditive Models." In *Sociological Methodology 1980,* ed. Karl F. Schuessler, 459–88. San Francisco: Jossey-Bass, 1979.

Tate, Richard L. "Limitations of Centering for Interactive Models." *Sociological Methods and Research* 13, no. 2 (1984): 251–71.

Tsebelis, George. *Veto Players: How Political Institutions Work.* New York: Russell Sage Foundation, 2002.

Western, Bruce. "Causal Heterogeneity in Comparative Research: A Bayesian Hierarchical Modelling Approach." *American Journal of Political Science* 42, no. 4 (1998): 1233–59.

Wolfinger, Raymond E., and Steven J. Rosenstone. *Who Votes?* New Haven: Yale University Press, 1980.

Zedeck, Sheldon. "Problems with the Use of 'Moderator' Variables." *Psychological Bulletin* 76, no. 4 (1971): 295–310.

INDEX

behavioral models, 1, 3, 8, 11, 13, 15, 103–4. *See also* social-welfare example; voter-turnout example
binary outcomes, 111–23

centering, 21–22, 49, 93–99
chained interaction, 39–42, 59–60, 73–74
Classical Linear Regression Model, assumptions, 17, 47, 83, 107–11, 124–29
cluster, 126–30
colinearity, 7, 77, 93–99, 113
conditional effects. *See* marginal effects
confidence intervals: hourglass shape, 64–65, 73–74, 84–85
   hypothesis tests, 47–49, 51–53, 61–78
   marginal effects, 61–78
   predicted values, 79–88
   predicted values, differences in, 88–91
Cox, Gary W. *See* presidential-candidates example
Cronbach, Lee J., 93–99
cumulative probability distribution function, 114, 119, 135

delta method, 119–23
derivative/first derivatives. *See* differentiation
deterministic (vs. stochastic) relationships, 5–6, 14, 16–18, 110, 123–30
dichotomous variables. *See* dummy variables, dependent and independent
differences in predicted values. *See* predicted values
differentiation, 20, 22–43, 50, 61–78, 88–91, 111–23, 134–35. *See also* marginal effects
direct vs. indirect effects, 19–22
discrete changes, effects of, 26–28, 88–91, 114, 119. *See also* dummy variables, independent; ordinal variables; predicted values
dummy variables, dependent. *See* binary outcomes
dummy variables, independent, 27, 28–29, 67–68, 74–75, 85–86, 88–89, 103–11

feasible generalized least squares, 110, 127–28, 129

feasible weighted least squares, 128, 129
F-tests, 45, 50–51, 54–60, 73, 108–10

government-duration example, 29–32, 34–37, 39–42, 44, 56–60, 69–78, 86–88, 91, 137
graphing: choosing values to graph, 61–66, 141–42
  marginal effects with confidence intervals, 61–87, 136–40
  predicted values with confidence intervals, 88–91, 141–44

heteroskedasticity, 17, 109–11, 124–30
hierarchical models, 5, 17, 123–30
hierarchical testing, 99–102
homoskedasticity, 17, 109–11, 124–30
Huber-White standard errors, 126, 128
hypothesis tests, 43–60
  directional, 45–50, 53–55, 57
  null hypothesis, defined, 45

independent vs. total effects, 19–22
institutions/institutional analysis, 2–3, 8–11. See also government-duration example; hierarchical models; presidential-candidates example
interactive arguments, 2–3, 8–12, 14–16

lincom (STATA command), 61, 138, 144, 146
linear-additive model, 1–2, 14, 20, 46
logarithmic transformations, 27, 32, 34–37, 58–59, 70–73, 87–88, 91, 135
logit/logistic. See binary outcomes

main vs. interactive effects, 19–21, 23, 43–44, 64
marginal effects, 22–43, 44–60, 61–78, 88–91, 111–23, 132, 136–41, 144, 146

confidence intervals, 61–87, 136–40
standard errors, 48, 52, 58–59, 119–22
mean-centering, mean-rescaling. See centering
mediating vs. moderating variables, 3
moderating vs. mediating variables, 3
multilevel models. See hierarchical models

nonlinear models, 5, 28, 111–23
nonlinear transformations. See logarithmic transformations; quadratic transformations; slope-shift models; spline models

omitted-variable bias, 46, 93–94, 99–102
ordinal variables, 26–27, 103, 115–16

pairwise interaction, 40–42, 74–78
path analysis, 3
pooled-sample vs. separate-sample estimation, 5, 103–11
postestimation commands, 48, 61, 81, 122–23, 138. See also lincom
predicted probabilities, 115, 118–19, 122–23
  differences in and standard errors around, 119, 122–23
predicted values, 25–26, 28–39, 61, 78–92, 141–44. See also predicted probabilities
  differences in and standard errors around, 25–26, 33–34, 36, 88–92
  confidence intervals, 78–92
  standard errors, 78–92, 141–44
presidential-candidates example, 13–14, 18–21, 23–26, 44, 51–54, 62–67, 81–85, 90–91
principal-agent models, 10–11, 13
probability density function, 114, 135
probit. See binary outcomes

quadratic transformations, 32–34, 35, 57, 69–71, 86, 91

random-coefficient models. *See* random-effects models
random-effects models, 5, 17, 123–30
robust standard errors, 126–30

separate-sample estimation vs. pooled-sample estimation, 5, 103–11
slope-shift models, 37–39
social-welfare example, 27–29, 44, 54–56, 67–68, 85–86, 106–8
spatial correlation, 11, 126
spline models, 37–39
standard errors: of coefficient estimates. *See* variance-covariance matrix, defined
  of differences in predicted values. *See* predicted values, differences in and standard errors around
  of marginal effects. *See* marginal effects, standard errors
  of predicted values. *See* predicted values, standard errors
STATA, 48, 61, 129, 136–46
statistical tests of interactive arguments, 43–60
stochastic vs. deterministic relationships, 5–6, 14, 16–18, 110, 123–30

tables: choosing values to list, 61–66, 141–42
  differences in predicted values and standard errors around, 88–91, 141
  marginal effects and standard errors around, 52–57, 136–40
  predicted values and standard errors around, 78–88, 141–44
temporal correlation, 11, 126
three-way interactions, 39–42, 59–60, 73–78
threshold model, 37–39
total vs. independent effects, 19–22
t-tests, 43–60, 72–73, 99–102, 108–11

variance: of differences in predicted values. *See* predicted values, differences in and standard errors around
  of forecast error, 47
  of a linear combination of constants and random variables, 47, 80–81
  of marginal effects. *See* marginal effects, standard errors
  of predicted values. *See* predicted values, standard errors
variance-covariance matrix
  defined, 48
  examples, 52, 55, 56, 121, 138–45
voter-turnout example, 115–22

White's robust standard errors, 125–30